T0331027

Graph Learning Techniques

This comprehensive guide addresses key challenges at the intersection of data science, graph learning, and privacy preservation.

It begins with foundational graph theory, covering essential definitions, concepts, and various types of graphs. The book bridges the gap between theory and application, equipping readers with the skills to translate theoretical knowledge into actionable solutions for complex problems. It includes practical insights into brain network analysis and the dynamics of COVID-19 spread. The guide provides a solid understanding of graphs by exploring different graph representations and the latest advancements in graph learning techniques. It focuses on diverse graph signals and offers a detailed review of state-of-the-art methodologies for analyzing these signals. A major emphasis is placed on privacy preservation, with comprehensive discussions on safeguarding sensitive information within graph structures. The book also looks forward, offering insights into emerging trends, potential challenges, and the evolving landscape of privacy-preserving graph learning.

This resource is a valuable reference for advance undergraduate and postgraduate students in courses related to Network Analysis, Privacy and Security in Data Analytics, and Graph Theory and Applications in Healthcare.

Baoling Shan is currently a lecturer at the University of Science and Technology Beijing, Beijing, China.

Xin Yuan is currently a senior research scientist at CSIRO, Sydney, NSW, Australia, and an adjunct senior lecturer at the University of New South Wales.

Wei Ni is a principal research scientist at CSIRO, Sydney, Australia, a Fellow of IEEE, a conjoint professor at the University of New South Wales, an adjunct professor at the University of Technology Sydney, and an honorary professor at Macquarie University.

Ren Ping Liu is a professor and the head of the Discipline of Network and Cybersecurity, University of Technology Sydney (UTS), Ultimo, NSW, Australia.

Eryk Dutkiewicz is currently the head of School of Electrical and Data Engineering at the University of Technology Sydney, Australia. He is a senior member of IEEE and his research interests cover 5G/6G and IoT networks.

Graph Learning Techniques

Baoling Shan, Xin Yuan, Wei Ni,
Ren Ping Liu and Eryk Dutkiewicz

CRC Press
Taylor & Francis Group
Boca Raton London New York

CRC Press is an imprint of the
Taylor & Francis Group, an **informa** business

Designed cover image: The Modern District Of Msheireb With Tram In The Downtown Of Doha High-Res Stock Photo - Getty Images

First edition published [2025]
by CRC Press
2385 NW Executive Center Drive, Suite 320, Boca Raton FL 33431

and by CRC Press
4 Park Square, Milton Park, Abingdon, Oxon, OX14 4RN

CRC Press is an imprint of Taylor & Francis Group, LLC

ISBN: 978-1-032-85113-6 (hbk)
ISBN: 978-1-032-85112-9 (pbk)
ISBN: 978-1-003-51661-3 (ebk)

DOI: 10.1201/9781003516613

Typeset in CMR10 font
by KnowledgeWorks Global Ltd.

Contents

Abstract

Graphs serve as a widely recognized representation of the network structure of interconnected data. They appear in various application domains, including social systems, ecosystems, biological networks, knowledge graphs, and information systems. As artificial intelligence technologies continue to advance, graph learning (i.e., machine learning applied to graphs) is attracting increasing interest from both researchers and practitioners. This approach has proven effective for numerous tasks, such as classification, link prediction, and matching, by utilizing machine learning algorithms to extract pertinent features from graphs. A critical challenge is to excavate graphs underlying observed signals because of non-convex problem structure and associated high computational requirements. On the other hand, latent graph structure and stimulus of graph data contain critical private information, such as brain disorders in functional magnetic resonance imaging data, and can be exploited to identify individuals. It is critical to perturb the latent information while maintaining the utility of the data, which, unfortunately, has never been addressed.

This book provides a comprehensive exploration of graph learning, covering fundamental concepts, state-of-the-art technologies, and emerging trends. It begins with an introduction to graph structures, including definitions, types, and basic concepts, followed by an overview of graph-structured data and graph learning techniques, particularly graph neural networks. Privacy considerations in graph learning are addressed, emphasizing privacy risks, differential privacy principles, and privacy-preserving techniques. The book then delves into existing technologies for graph learning, such as statistical methods, learning graphs from smooth and stationary signals, and band-limited graph processes. Detailed discussions on graph extraction, topology learning, and the application of the Graph Fourier Transform (GFT) for band-limited data are included. Special focus is given to specific applications, including brain signal analysis and COVID-19 spread analysis, highlighting graph inference methods and result assessments. The book also explores methods for preserving the privacy of latent information in graph-structured data through obfuscation and perturbation techniques. Concluding with future directions and challenges, the book outlines emerging trends in privacy and graph learning, addressing scalability issues and proposing future research directions. This comprehensive survey aims to serve as a valuable resource for researchers and practitioners in the field of graph learning.

List of Figures

List of Tables

Contributors

Eryk Dutkiewicz

University of Technology Sydney

Sydney, Australia

Ren Ping Liu

University of Technology Sydney

Sydney, Australia

Wei Ni

Data61, CSIRO

Sydney, Australia

Baoling Shan

National School of Elite Engineering

University of Science and Technology
  Beijing

Beijing, China

Xin Yuan

Data61, CSIRO

Sydney, Australia

1

Introduction

This chapter introduces the foundational concepts of graph theory and graph learning. It covers the definitions, types, and representations of graphs, and explores the nature of graph-structured data, including graph signals and their transformations. The chapter discusses graph signal processing and representation learning, emphasizing the importance of semi-supervised and unsupervised learning methods. It also examines the interplay between graph learning and graph neural networks. Practical applications in brain networks and COVID-19 analysis illustrate the significance of graph learning. The chapter ends with an overview of the book's structure, preparing readers for the detailed discussions ahead.

1.1   What Is a Graph?

Graphs, also known as networks, can be derived from various real-world relationships among numerous entities. Common types of graphs are frequently used to represent different kinds of relationships, including social networks, biological networks, patent networks, traffic networks, citation networks, and communication networks. A graph typically consists of two sets: A vertex set and an edge set. Vertices represent the entities within the graph, while edges represent the relationships between these entities.

1.1.1   Definition and Basic Concepts

A graph $\mathcal{G}$ is a mathematical structure used to model pairwise relationships between objects. Define $\mathcal{G} = (\mathcal{V}, \mathcal{E}, \mathbf{W})$, where $\mathcal{V} = \{1, \cdots, N\}$ is the set of N vertices and $\mathcal{E} \subseteq \mathcal{V} \times \mathcal{V}$ is the set of edges. $\mathbf{W} \in \mathbb{R}^{N \times N}$ is the weighted adjacency matrix of $\mathcal{G}$, which indicates to what extent two nodes are correlated. A line between the vertices i and j indicates the existence of an edge between vertices i and j, that is, $(i, j) \in \mathcal{E}$. Examples of graph topologies with $N = 7$ vertices, with $\mathcal{V} = 0, 1, 2, 3, 4, 5, 6$ are presented in Fig. 1.1, along with the corresponding edges. The vertices are usually depicted as points (circles), and the edges are lines that connect the vertices.

DOI: 10.1201/9781003516613-1

1.1.2 Types of Graphs (directed, undirected, weighted)

Regarding the directionality of vertex connections, a graph can be directed and undirected, as illustrated in Figs. 1.1 and 1.2, respectively.

Directed Graphs (Digraphs): In directed graphs, each edge has a direction, indicating a one-way relationship from one vertex to another. For example, the directed graph in Fig. 1.2 can be described as

$$\mathcal{V} = (0, 1, 2, 3, 4, 5, 6), \tag{1.1}$$

$$\mathcal{E} = \{(0, 6), (1, 0), (1, 2), (2, 3), (2, 6), (6, 2), \\ (3, 5), (5, 3), (4, 3), (5, 4), (5, 0), (5, 6)\}. \tag{1.2}$$

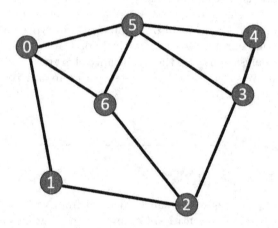

FIGURE 1.1: Undirected graph.

Undirected Graphs: In undirected graphs, edges do not have directions, indicating bidirectional relationships between the vertices. With the difference to the directed graph, if $(i, j) \in \mathcal{E}$, then $(j, i) \in \mathcal{E}$.

Weighted Graphs: Weighted graphs have edges assigned with weights, representing the cost, length, or capacity of the connection between the vertices.

1.1.3 Graph Representations

For a given set of vertices and edges, a graph can be formally represented by its adjacency matrix $\mathbf{A}$, which describes the vertex connectivity; for N vertices, the adjacent matrix $\mathbf{A}$ is an $N \times N$ matrix. The elements $\mathbf{A}_{ij}$ of the adjacency matrix $\mathbf{A}$ assume values $\mathbf{A}_{ij} \in \mathcal{E}$. The value $\mathbf{A}_{ij} = 0$ is assigned if the vertices i and j are not connected with an edge, and $\mathbf{A}_{ij} = 0$ if these

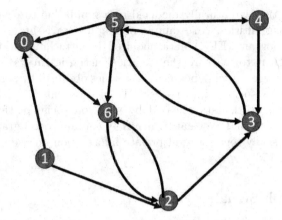

FIGURE 1.2: Directed graph.

vertices are connected, that is,

$$\mathbf{A} = \begin{cases} 1, & \text{if } (i,j) \in \mathcal{E}; \\ 0, & \text{if } (i,j) \notin \mathcal{E}. \end{cases} \tag{1.3}$$

A nonzero element in the weight matrix $\mathbf{W}, W_{ij} \in \mathbf{W}$, designates both an edge between the vertices i and j and the corresponding weight. $\mathbf{W} \in \mathbb{R}^{N \times N}$ is the weighted adjacency matrix of $\mathcal{G}$, which indicates to what extent two nodes are correlated. $W_{ij} = W_{ji} \neq 0$ for any $(i,j) \in \mathcal{E}$.

Another important descriptor of graph connectivity is the combination graph Laplacian matrix, $\mathbf{L}$, which combines the weight matrix and the degree matrix

$$\mathbf{L} = \mathbf{D} - \mathbf{W}, \tag{1.4}$$

where $\mathbf{D} \triangleq \text{diag}(\mathbf{W1})$ is the diagonal matrix containing the degrees of vertices of the graph at its diagonal, and $D_{ij} = 0, i \neq j$. $\mathbf{1}$ is an all-one vector. For an undirected graph, the Laplacian matrix is symmetric, that is, $\mathbf{L} = \mathbf{L}^T$. The space of the combination graph Laplacian matrices is defined as

$$\mathcal{L} = \{\mathbf{L} \in \mathbb{R}^{N \times N} | L_{ij} = L_{ji} \leq 0, i \neq j\}. \tag{1.5}$$

1.2 Graph-Structured Data

An important type of data is graph data, which is the data with latent graph structures (or in other words, a set of concurrent time series with underlying correlations) observed extensively in the areas of physics, biology, transportation, energy, engineering, and social science [128]. Graph-structured data

refers to data that is naturally modeled as a graph due to its inherent relationships among entities. An example of graph data is brain data, such as
electroencephalogram (EEG) data acquired by attaching electrodes to the
scalp of a patient to measure the electrical activities in its brain [138], as
well as blood-oxygen-level-dependent time series obtained by performing functional magnetic resonance imaging (fMRI) on the brain [61]. More examples
include social network data released by users on social platforms, such as
Facebook, Twitter, and WeChat, which may contain social graphs and large
amounts of possibly sensitive and private information on users and their interactions [83, 158].

1.2.1   Graph Signal

A graph signal is a function that assigns a scalar or vector to each vertex. For
simplicity, consider $\mathbf{x} : \mathcal{V} \rightarrow \mathbb{R}$, representing the intensity on each vertex of a
mesh. Graph signals are denoted as $\mathbf{x} \in \mathbb{R}^N$, where y_i represents the signal
value at the i-th vertex.

1.2.2   Graph Fourier Transform (GFT)

GFT decomposes a graph signal into orthonormal components that describe
various modes of variation according to the graph topology encoded in the
Laplacian matrix $\mathbf{L}$ (or another graph-shift operator dictated by the application). The GFT enables the representation of a graph signal in two distinct
domains: The vertex domain, consisting of the nodes in $\mathbf{U}$, and the graph
frequency domain, defined by the spectral basis of $\mathcal{G}$. This dual representation allows for the manipulation of signals in the frequency domain, enabling
different levels of interaction between neighboring nodes in the network. The
GFT captures the concept of signal variability across the graph, analogous to
the notion of frequency in the Fourier analysis of temporal signals.

To elaborate on this concept, consider the eigenvector decomposition of the
combinatorial graph Laplacian $\mathbf{L}$ to define the GFT and the associated notion
of graph frequencies. Let $\mathbf{\Lambda} = \text{diag}(\lambda_1, \ldots, \lambda_N)$ denote the diagonal matrix
of nonnegative Laplacian eigenvalues and $\mathbf{U} = [\mathbf{u}_1, \ldots, \mathbf{u}_N]$ the orthonormal
matrix of eigenvectors, one can decompose the symmetric graph Laplacian as
$\mathbf{L} = \mathbf{U}\mathbf{\Lambda}\mathbf{U}^T$.

Definition 1.2.1 *(**GFT**): The GFT of $\mathbf{x}$ with respect to the combinatorial
graph Laplacian L is the signal $\hat{\mathbf{x}} = [\hat{x}_1, \ldots, \hat{x}_N]^T$ defined as $\hat{\mathbf{x}} = \mathbf{U}^T x$. The
inverse GFT (iGFT) of $\hat{\mathbf{x}}$ is given by $\mathbf{x} = \mathbf{U}\hat{\mathbf{x}}$, which is a proper inverse due
to the orthogonality of $\mathbf{U}$.*

The iGFT formula $\mathbf{x} = \mathbf{U}\hat{\mathbf{x}} = \sum_{k=1}^{N} \hat{x}_k \mathbf{u}_k$ allows for synthesizing x as a sum
of orthogonal frequency components $\mathbf{u}_k$. The contribution of each $\mathbf{u}_k$ to the
signal $\mathbf{x}$ is given by the GFT coefficient $\hat{x}_k$.

The IGFT reconstructs the graph signal from its frequency content by combining graph frequency components weighted by the coefficients of the signal's GFT. With an appropriately constructed graph that captures the signal structure well, the GFT will lead to a compact representation of the graph signal in the spectral domain, which is beneficial for geometric data processing such as reconstruction and compression.

1.2.3 Sampling and Reconstruction of Graph Signals

Suppose that the coefficients of a graph signal $\mathbf{x} \in \mathbb{R}^N$ are sampled to produce a signal $\mathbf{x}_\mathcal{M}, \mathcal{M} \in \mathbb{R}^M (M < N)$, where $\mathcal{M} = (\mathcal{M}_0, \cdots, \mathcal{M}_{M-1})$ denotes the sequence of sampled indices, and $\mathcal{M}_i \in \{0, 1, \cdots, N-1\}$. The signal $\mathbf{x}_\mathcal{M}$ is then interpolated to get $\mathbf{x}' \in \mathbb{R}^N$, which recovers either exactly or approximately. The sampling operator Ψ is a linear mapping from $\mathbb{R}^N$ to $\mathbb{R}^M$, defined as

$$\Psi_{i,j} = \begin{cases} 1, & j = \mathcal{M}_i; \\ 0, & \text{otherwise}, \end{cases} \tag{1.6}$$

and the interpolation operator Φ is a linear mapping from $\mathbb{R}^M$ to $\mathbb{R}^N$.

$$\textbf{\textit{Sampling}} : \mathbf{x}_\mathcal{M} = \Psi\mathbf{x}; \tag{1.7}$$

$$\textbf{\textit{Interpolation}} : \mathbf{x}' = \Phi\mathbf{x}_\mathcal{M} = \Phi\Psi\mathbf{x},$$

where $\mathbf{x}' \in \mathbb{R}^N$ recovers either exactly or approximately.

1.2.4 Smooth and Band-Limited Graph Signals

Definition: A graph signal is smooth if the signal values associated with the two end vertices of edges with large weights in the graph tend to be similar.

The smoothness of graph signal $\mathbf{x}$ on graph $\mathcal{G}$ with positive edge weights can be quantified by the graph Laplacian quadratic form

$$\mathbf{x}^T \mathbf{L} \mathbf{x} = \frac{1}{2} \sum_{i,j} W_{ij} \|y_i - y_j\|^2, \tag{1.8}$$

where W_{ij} denotes the weight of the edge between nodes i and j. In other words, graph signal $\mathbf{x}$ is considered smooth if each sample value y_i on the vertex i is similar to the one y_j on the neighboring vertex j with large W_{ij}.

Definition: A graph signal is called bandlimited when there exists a $K \in \{0, \cdots, N-1\}$ such that its GFT satisfies

$$\mathbf{s}_k = \mathbf{U}^T \mathbf{x}_k = 0, \text{for all } k \geq K. \tag{1.9}$$

The smallest such is called the bandwidth of $\mathbf{x}$. Such signals can be sampled and reconstructed more efficiently, as their information content is concentrated in a smaller set of frequencies.

1.3 Graph Learning

The graph topology captures the instinct relationships of unstructured data encoded on different entries of the graph and graph learning is an efficient technique used to uncover the latent graph topologies of data [45]. Classical graph topology inference methods, such as graph lasso [60] and covariance selection [40], estimated the covariance matrices of graph signals. More recent graph learning techniques enforced smoothness to graph signals prior to topology inference [44, 86, 62, 135, 28, 142, 131]. In other words, it has been typically assumed that the frequency-domain representations of graph signals have unlimited bandwidths, e.g., for mathematical tractability.

With the latent graph topologies, the captured data can be transformed into the frequency domain, where data can be effectively processed. This is critical to recover missing data in part of a network [21], or verify the authenticity of data [58]. For the sake of generality, it is reasonable to consider band-limited graph signals; i.e., the frequency-domain representation of the signals can have a finite bandwidth. The band-limitedness is observed in practice, e.g., fMRI data in brain networks [72]. The widely studied smooth graph signals with unlimited frequency-domain bandwidths [44] can be viewed as a special case of band-limited graph signals.

Generally speaking, graph learning is the technique used to uncover the latent graph topologies of data [45]. It refers to machine learning on graphs. Graph learning methods map the features of a graph to feature vectors with the same dimensions in the embedding space. A graph learning model or algorithm directly converts the graph data into the output of the graph learning architecture without projecting the graph into a low-dimensional space. Most graph learning methods are based on or generalized from deep learning techniques because deep learning techniques can encode and represent graph data into vectors. The output vectors of graph learning are in continuous space. The target of graph learning is to extract the desired features of a graph. Consequently, graph learning is a more powerful and meaningful technique for graph analysis.

Existing graph learning methods cannot accurately and efficiently infer the graph topology (i.e., graph Laplacian [177]) of band-limited graph signals due to difficulties in joint estimation of both the frequency-domain representation and the GFT basis converting captured data to the frequency domain.

1.3.1 Graph Signal Processing

Graph signal processing (GSP) was introduced a decade ago in the influential works of [67, 146, 136]. Since then, GSP-related issues have garnered substantial interest, not only in the signal processing (SP) community [149] but also in machine learning (ML) forums, where research in graph-based learning has grown considerably [17]. Graph signals are ideal for modeling data linked to a set where 1) the elements belong to the same category (e.g., regions of the cerebral cortex, members of a social network, weather stations); 2) there is a relationship (physical or functional) of proximity, influence, or association among these elements; and 3) the strength of such relationships varies among pairs of elements.

In some cases, the supporting graph represents a physical, technological, social, information, or biological network with observable links. In other scenarios, the graph is implicit, reflecting dependencies or similarities among nodes, with links inferred from the data. Thus, GSP is a comprehensive framework that extends classical SP methods, tools, and algorithms to modern technological domains, including social, transportation, communication, and brain networks; recommender systems; financial engineering; distributed control; and learning.

GSP [122, 137] provides a powerful framework for processing and analyzing signals that are associated with graph structures. It extends traditional signal processing techniques to signals that are defined on graphs, enabling the analysis of various data types and applications [132]. GSP encompasses a range of methods and algorithms that operate on graph-structured data. These include graph signal filtering [23, 31], which involves modifying or extracting specific components of graph signals based on their spectral properties. Graph signal compression [9] aims to reduce the size of graph signals while preserving important information. Graph signal sampling [27, 154] focuses on selecting a subset of graph nodes to represent the entire signal efficiently. Furthermore, GSP encompasses techniques for recovering data on irregular graph domains [22].

The main objective of GSP is to uncover the underlying dependencies, physical proximity, or other properties within datasets that are indexed by vertices in a graph [154]. By considering the graph structure, GSP enables the exploration and analysis of relationships and patterns in the data, leading to insights and interpretations that may not be apparent in traditional signal processing approaches [146, 42, 116]. However, one challenge in applying GSP is that the graph structure is not always readily available [112]. In earlier studies, the underlying graph topology was assumed to be known, and GSP was used to analyze how graph signals can affect the algebraic characteristics of the underlying graphs [136]. More recently, efforts have been made to reconstruct the network structure from the data itself, allowing for the expression of complex relations in real systems [136]. GSP has gained increasing interest and has been applied in various fields, including social networks [84], neuroscience [104], image processing [161, 76], sensor networks [82, 54, 95], and communication systems [73, 105]. It provides a powerful framework for understanding and exploiting the rich information encoded in graph-structured data, enabling advancements in data analysis, pattern recognition, and decision-making processes.

1.3.2   Overview of Graph Representation Learning

Traditional methods [40, 60] estimated the covariance matrix to capture the linear/nonlinear and symmetric pairwise or directional dependency among vertex-indexed signals. Such methods only captured pairwise correlations and did not reflect the causality of latent network structures [36].

Later, graph learning techniques were developed to infer the topology from observations, where graph signals were assumed to be smooth; in other words, their frequency-domain representations have unlimited bandwidths. These techniques are not suitable for band-limited signals. Dong *et al.* [45] outlined techniques to solve graph learning problems for globally and locally smooth models, and summarized the potential benefit of GSP-based graph inference methods in many theoretical and practical applications, such as image coding, brain functional connectivity analysis, and meteorology analysis. Chepuri *et al.* [28] studied the learning of a sparse graph to explain observed data under a smoothness prior adequately. They provided an AO algorithm and a one-step convex relaxation-based solution by modeling the learning problem as a sparse edge sampling function. Dong *et al.* [44] inferred the graph Laplacian by minimizing the variations of the smooth signals based on the *a-priori* information about the structure. This method was recently used in [58] to reconstruct missing air pollution data and experimentally verified. Kalofolias *et al.* [86] extended this idea by constructing graph learning as a weighted sparsity problem, and learned a valid structure represented by the adjacent weighted matrix using primal-dual optimization. The method of [86] was recently used in [62] to study the functional connectivity of brain networks from fMRI time series. Saboksayr *et al.* [135] extended the method of [86] to support general multi-class smooth signals. Egilmez *et al.* [52] proposed a block-coordinate descent-based algorithm to learn the graph Laplacian. The original problem was decomposed into subproblems based on the Laplacian estimation's structural constraints and optimality conditions. The subproblems were solved in an alternating manner at each iteration. Egilmez's algorithm was applied in [51] for video compression with improvements over the widely used Karhunen-Loeve transform.

Some recent works learned topologies generated by diffusion processes on graphs, typically under the assumption of stationarity; i.e., the graph diffusion operator (e.g., the adjacency or Laplacian matrix) has the same eigenvectors as the covariance matrix of observed signals. Segarra *et al.* [142] estimated a new graph shift operator by minimizing the ℓ_1-norm of the operator based on the complete or partial knowledge of the eigenvectors of the covariance. They used independent samples of signals to evaluate the eigenvectors and then estimated the eigenvalues given the eigenvectors. The obtained graph shift operator supported band-limited signals when partial knowledge about the eigenvectors was available, and the observed signal samples were band-limited. Pasdeloup *et al.* [131] studied the case where signals were independent and identically distributed (i.i.d.) and observed after diffusion on a graph. It was verified that the set of graphs has impacts on the eigenvectors of the covariance matrix used to recover the graph topology.

A few existing studies [140, 80] have attempted to learn the latent graphs of band-limited signals. Their accuracy and efficiency were penalized by their AO-based approximate solvers. Specifically, Sardellitti *et al.* [140] discovered a block sparse representation of band-limited graph signals and developed a strategy to associate a graph with observed band-limited signals. The strategy started by learning an orthonormal sparsifying transform based on the AO. The resulting problem of the graph estimation became convex. The strategy recovered the graph Laplacian with convex optimization techniques.

1.3.3 Graph Learning in Semi-Supervised and Unsupervised Settings

Graph learning in semi-supervised and unsupervised settings is a crucial area of research that focuses on effectively utilizing graph-structured data to make predictions and extract useful representations.

- *Semi-supervised graph learning:* Semi-supervised graph learning models often rely on the assumption that connected nodes share similar labels, capitalizing on the graph's topology to propagate information. Semi-supervised graph learning [148] leverages both labeled and unlabeled nodes within a graph to improve learning outcomes. This approach is particularly effective when labeled data is scarce, but the graph structure provides rich information.

  Key techniques in this domain include Graph Convolutional Networks (GCNs), which generalize convolutional operations to graph data by aggregating information from a node's neighbors. Another prominent method is Graph Attention Networks (GATs), which assign varying importance to different neighbors using attention mechanisms, enhancing the focus on relevant connections.

  Label propagation algorithms iteratively spread labels through the graph, smoothing labels based on connectivity. Graph Sample and Aggregate (GraphSAGE) is another method that samples and aggregates features from a fixed-size neighborhood, making it scalable to large graphs and applicable to unseen nodes. These techniques are widely used for node classification tasks, such as predicting user interests in social networks or identifying categories in citation networks.

The effectiveness of semi-supervised graph learning hinges on the model's ability to capture and utilize the underlying graph structure, making it a powerful tool for applications requiring nuanced understanding from limited labeled data.

- ***Unsupervised graph learning:*** Unsupervised graph learning focuses on extracting meaningful representations from graph-structured data without any labeled nodes. This approach is crucial for tasks where labeled data is unavailable, allowing the discovery of latent patterns purely from the graph structure. Key techniques in unsupervised graph learning include Graph Autoencoders (GAEs) [99, 98], which encode graph data into a lower-dimensional space and decode it to reconstruct the graph, capturing essential structural information.

Variational Graph Autoencoders (VGAEs) extend GAEs by introducing probabilistic elements and learning distributions over node embeddings rather than fixed representations [97]. Methods like DeepWalk treat random walks on graphs as sentences, using techniques from natural language processing to learn node embeddings. Node2Vec enhances this approach by using biased random walks, enabling it to capture diverse node relationships such as homophily and structural equivalence.

GraphGAN integrates generative adversarial networks with graph learning, where a generator creates fake edges and a discriminator distinguishes them from real edges, refining both in the process. These techniques are employed in various applications, including node clustering, graph visualization, and anomaly detection. A significant challenge in unsupervised graph learning is ensuring scalability to large graphs, which is often addressed through efficient sampling and parallel processing methods.

Heterogeneous graphs containing different types of nodes and edges add complexity and necessitate advanced models like heterogeneous graph neural networks. Dynamic graphs, which evolve over time, require specialized methods to capture temporal changes. Despite these challenges, unsupervised graph learning provides powerful tools for uncovering hidden patterns and structures in graph data, making it invaluable for exploratory data analysis and scenarios where labeling is impractical.

1.4 Graph Learning and Graph Neural Networks

In multiple real-world scenarios, learning problems rely on non-Euclidean structures such as graphs, point clouds, or manifolds, rendering conventional deep learning models and approaches unsuitable. To mitigate this issue, graph neural networks (GNNs) were proposed to leverage the full underlying structure of the dataset and maximize learning capacity by directly learning from the graph. GNNs are an effective framework for the representation learning of graphs. GNNs follow a neighborhood aggregation scheme, where the representation vector of a node is computed by recursively aggregating and transforming representation vectors of its neighboring nodes [169]. GNNs are a subset of graph learning methods specifically designed to process and learn from graph-structured data using neural networks. They extend neural network operations to graphs, capturing the dependencies between nodes through their connections [185, 167]. GNNs are deep learning-based methods that operate on the graph domain. Due to its convincing performance, GNN has recently become a widely applied graph analysis method.

Convolutional Neural Networks (CNNs) [25] have the ability to extract multi-scale localized spatial features and compose them to construct highly expressive representations, which led to breakthroughs in almost all machine learning areas and started the new era of deep learning [92]. The keys of CNNs are local connection, shared weights, and the use of multiple layers.

1.5 Graph Learning for Brain Networks Analysis

Network neuroscience plays a crucial role in advancing our understanding of the structure and function of the human brain. It adopts a network perspective by considering the brain as a complex system composed of multiple regions of interest (ROIs), often referred to as brain network nodes [115]. Graph theory [125] has been a crucial tool to analyze complex brain networks, and revealed several non-trivial features of brains, such as modularity and small-worldness, by studying the fMRI time series (i.e., blood-oxygen-level dependent time series) amongst the ROIs in a brain [115].

Traditional brain graph learning methods [40, 60, 63] estimated the covariance matrix to capture the linear/nonlinear and symmetric pairwise or directional dependency among vertex-indexed observed signals. Pearson's correlation has been one of the most common methods for measuring the pairwise functional relationships between brain regions. However, Pearson's correlation focuses on strong direct marginal correlations of the fMRI time series between two brain regions, and overlooks the latent network effects of other brain regions. Albert *et al.* [4] used partial correlations to measure interactions between any two ROIs. The partial correlation quantifies the dependency between two ROIs by regressing out the other ROIs, leading to difficulties in the suppression of the confounding effect from the other ROIs. Those methods only captured pairwise correlations and cannot reflect the accurate causality in the latent network structures of brains [20, 36, 110]. Sparse inverse covariance estimation (SICE) [178] is another popular technique for measuring the intensity of the most significant direct connection between ROIs. SICE is a principled partial correlation algorithm. It tends to evaluate the sparsest reconstructive coefficient of each ROI and capture only local structures, rather than a representation of the global structure.

Recent graph learning techniques have attempted to address the limitations of SICE by incorporating both local and global features to establish a representation of the entire graph [3]. These techniques deal with specific properties of observed signals, such as smoothness [44, 86] and stationarity [142], but are unsuited for band-limited brain signals. On the other hand, fMRI-based brain data exhibit a distinct characteristic of band-limitedness and smoothness due to their underlying physiological properties [80]. They are different from typically considered (band-unlimited) smooth graph signals. In other words, the signals are sparse in the frequency domain. As a result, existing graph learning algorithms developed under the assumptions of smoothness and stationarity cannot readily apply to fMRI-based brain signals.

The band-limitedness is a widely observed property of fMRI-based brain data [140, 72]. The observed signals can be sparse in the frequency domain. With the inferred graph topology, it is possible to recover the graph signals throughout the entire brain network by only observing part of the signals. Sardellitti *et al.* [139] discovered a block sparse representation of general graph data by enforcing the band-limitedness of observed signals. Instead of assuming any diffusion process over a graph, a strategy was developed to relate a graph to the received band-limited signals. The strategy started by estimating an orthonormal sparsifying transform based on AO. The resulting problem of graph estimation was convex. The graph Laplacian matrix was then recovered using convex optimization techniques. Humbert *et al.* [79] considered graph learning from multivariate signals with both smoothness and band-limitedness. A three-step AO-based algorithm was developed to leverage manifold gradient descent and linear programming to learn the Laplacian of the graph.

GCNs [26] have also been employed in more recent approaches for learning brain networks through various methods. Based on Pearson's correlation matrix as the node features, Zhao *et al.* [181] developed a novel dynamic GCN to distinguish ADHD patients from health control for a better understanding of ADHD-associated brain dysfunctions. Li *et al.* [107] developed a BrainGNN framework based on a graph neural network by using the topological and functional information of fMRI for classification tasks. Zhou *et al.* [184] designed an interpretable GCN model to identify and classify Alzheimer's disease and quantify the discriminative features of the brain connectivity patterns. Table 1.1 collates the relevant studies to learn brain networks.

Table 1.1: The state of the art in graph learning techniques for brain signal analysis

Method	Description
Pearson's correlation	This algorithm measures the pairwise functional relationships between brain regions with a focus on strong direct marginal correlations of the fMRI time series between two brain regions, and overlooks the latent network effects of other brain regions.
Partial correlation	This algorithm measures the dependency between two nodes by regressing out the remaining ones, leading to difficulties in removing the confounding effect from other nodes.
SICE [178]	As the principled method for partial correlation, SICE tends to evaluate the sparsest reconstructive coefficient of each ROI and capture only local structures, rather than a representation of the global structure.
Dong's algorithm [44]	This algorithm is an alternating minimization algorithm that infers the assumption of the smoothness of the signals.
Kalofolias' algorithm [86]	This algorithm is a scalable primal-dual algorithm that learns the topological structures represented by the adjacent weighted matrix of graphs.
Sardellitti's Total Variation (TV) algorithm [139]	This algorithm is a two-step strategy consisting of (a) learning the orthonormal sparsifying transform from data via AO and (b) then recovering the Laplacian matrix from the sparsifying transform via convex optimization.

Sardellitti's Estimated-Signal-Aid (ESA) algorithm [140]	Different from Sardellitti's TV graph learning algorithm, this two-step strategy recovers the Laplacian matrix from the sparsifying transform and GFT coefficients by using convex optimization in Step 2.
Humbert's algorithm [79]	This is an AO-based algorithm with three alternating steps relying on standard minimization methods, i.e., manifold gradient descent and linear programming. It learns graphs from multivariate signals with smoothness and band-limitedness.

1.6 Graph Learning for COVID-19 Analysis

The COVID-19 pandemic is exacerbating global health, economic, and social challenges. As of April 2022, Europe recorded 192.09 million confirmed cases and more than 2 million deaths. Since the pandemic outbreak, researchers from various fields have extensively investigated the spread of the disease. The complex network theory based on a pair-wise configuration has been widely utilized for modeling the topological connectivity of the COVID-19 data on a global perspective [10, 85, 153]. Azad *et al.* [10] tracked the spread of COVID-19 by utilizing the network analysis in India based on the travel history of infected patients and revealed that international travel played a key role in the pandemic outbreak in a country. Jo *et al.* [85] developed an infected network using the contact tracing information of confirmed cases, and found that governmental measures had a strong impact on the COVID-19 spread network in Seoul. Through modeling tourism mobility as a complex network, Tsiotas *et al.* [153] created a multidimensional framework to understand the COVID-19 spread across countries. Chu *et al.* [29] constructed an air travel network structure to visualize the connectedness and evolution of the pandemic. Travel subnetworks were formed by aggregating airport data at the national level and adding it to a matrix capturing the flight recurrences between countries. Using a similar conceptualization, they also developed a pandemic space approach [30] that uses the historical correlation of confirmed cases to locate the connections between different countries. By integrating Bayesian parameter inference with a Watts–Strogatz small-world network epidemiological model, Syga *et al.* [151] inferred a time-varying COVID-19 transmission network in Germany. It was shown that government interventions reduced random contacts and transmission probabilities.

Numerous methods have been designed to infer the pandemic's time-dependent transmission network, compared to previous works on network-based models. For instance, the correlation coefficients were exploited to capture the linear/nonlinear and symmetric pairwise matrix between different regions [118, 147, 130, 123]. So *et al.* [147] constructed dynamic pandemic networks over time for 164 countries to forecast and assess the risk of the pandemic using network statistics. The connections in the networks were established based on the relationships of changes in the count of reported cases between the two regions. Pan *et al.* [130] used the Pearson correlation coefficient, time-lagged cross-correlation, and dynamic time wrapping to examine interactions in the evolution of pandemics across the different states of the United States. McMahon *et al.* [118] examined the spatial correlations of new active cases across different states in the United States and assessed their magnitude over time. Their results showed stronger correlations between urban areas compared to rural areas, revealing that the pandemic spread was largely driven by travel between cities. Using spatio-temporal correlation, Aral *et al.* [123] identified distinct spatial clusters and spatial associations among COVID-19 cases in Turkey, revealing that spatial analysis helped explain the spread of the disease.

Alguliyev *et al.* [5] created a conceptual graph model by taking into account various epidemiological traits of COVID-19, such as social distance, the period of contact with an infected individual, and demographic characteristics based on location, thereby enabling a visual representation of virus propagation. This helps determine undetected cases of infection. Ieracitano *et al.* [81] adopted a deep learning technique based on fuzzy logic to create a classification system for the early identification of COVID-19 cases utilizing portable chest X-ray (CXR) images. Absar *et al.* [2] developed a computer-assisted system for the automatic classification of CXR images of COVID-19 utilizing the Support Vector Machine (SVM) to enable fast diagnosis of COVID-19.

1.7 Book Organization

The structure of the book is organized as follows.

Chapter 1 introduces the essential concepts of graph theory, including definitions, types, and representations of graphs. It discusses graph-structured data, covering Graph Fourier Transform, and techniques for sampling and reconstructing graph signals. The chapter also provides an overview of graph learning, both semi-supervised and unsupervised, and connects these concepts to the use of Graph Neural Networks for various applications.

Chapter 2 addresses privacy considerations in graph and graph learning, highlighting the challenges and risks associated with graph data and learning graph structures. It explains the principles of differential privacy and its application to graphs and provides a formal framework for privacy analysis. The chapter also discusses implementing differential privacy in graph learning through techniques like privacy-preserving GFT, private graph filtering, and noise addition for signal privacy.

Chapter 3 reviews existing technologies in graph learning, including statistical methods like Correlation Networks and Graphical Lasso for inferring graph structures from data, focusing on pairwise relationships and sparsity constraints. It also discusses learning graphs from observations of smooth signals and specialized processes such as stationary and band-limited graph processes, highlighting methodologies tailored to capture these characteristics of graph-structured data.

Chapter 4 proposes a new graph learning technique, which learns weighted and undirected graph topologies, more specifically, the graph Laplacian matrices, from band-limited brain signals.

Chapter 5 presents a new graph learning technique for general band-limited graph signals, which learns the graph topology (i.e., the graph Laplacian matrix) of observed bandlimited graph signals to overcome the high computational complexity of the graph learning method proposed in Chapter 4.

Chapter 6 explores the application of graph topology learning to brain signals, specifically focusing on ADHD data. It introduces the study's materials and system model, discusses the process of graph inference, assesses the methods used, presents the results, and concludes with implications for understanding brain connectivity in ADHD.

Chapter 7 presents a new graph-learning technique to accurately infer the graph structure of COVID-19 data, helping to reveal the correlation of pandemic dynamics among different countries and identify influential countries for pandemic response analysis.

Chapter 8 presents a new approach to protect the privacy of the latent information underlying graph-structured data (e.g., the graph structure and the stimulus underlying the observed graph-structured data) while minimizing the perturbations on the observed graph-structured data to maintain the utility of the data.

Chapter 9 discusses future directions and challenges in the intersection of privacy and graph learning.

2

Privacy Considerations in Graph and Graph Learning

Privacy is a significant concern for some graph-structured data, for example, brain network data obtained by fMRI [61]. The graph-structured brain data can be held by the Department of Neurology in a hospital and shared with and used by other departments or clinics for big data analytics (e.g., detecting or modeling changes in blood flow that occur with brain activity) or educational purposes. On the one hand, the latent graph structures within brain network data could unveil personal health conditions, such as ADHD and AD, when subjected to graph interference attacks employing graph learning techniques [142, 140, 80]. These conditions could potentially expose patient identities [141, 127]. The latent stimuli within graph-structured data, serving as inputs to the latent graphs and contributing to the observed output, also constitute private information [116]. Bandwidth and waveform characteristics of these stimuli could serve as unique identifiers for individuals.

This chapter delves into privacy considerations in graph learning, highlighting the privacy risks, the sensitivity of graph signals, and threats involved in learning graph structures. It covers the basics of differential privacy, including its principles, application to graphs, and a formal framework for privacy analysis. This chapter also explores implementing differential privacy.

2.1 Introduction to Privacy Challenges in Graph Learning

Preserving the privacy of the graph-structured data, more explicitly, the privacy of the latent graph structures and stimuli underlying the data, has never been addressed in the literature. Existing studies have been dedicated to the privacy of graphs (as opposed to the graph-structured data), e.g., through k-anonymity [175], node or edge perturbation [108], and graph differential privacy (DP) [164, 109, 113]. The results of the studies cannot apply to the graph-structured data, i.e., a set of time series with correlations. In light of the notion of DP [48], a possible solution to obfuscating graph data is to add random noises to the graph data. However, the extent to which this DP-based obfuscation can effectively preserve the privacy of the latent graphs and stimuli is unclear.

Numerous privacy-preserving techniques for graph have been developed, e.g., k-anonymity [175], node and edge perturbation [69], and graph DP [48]. These methods do not apply to graph data, i.e., the three-dimensional data with latent graph structures (or, in other words, a set of concurrent time series), such as the brain signals. The methods would require the latent graphs to be extracted from the data, and separately obfuscated. The usefulness or effectiveness of graph data would degrade.

Yuan *et al.* [175] proposed an anonymization method that adds noisy nodes to the original graph to protect structural information. k-degree-l-diversity anonymity was achieved, where k measures the number of vertices with the same degrees, and l indicates the distinct labels. Ding *et al.* [43] developed a privacy-preserving framework to anonymize graphs, which defined an information loss matrix for graph datasets based on a k-decomposition method. k-anonymity was performed on isomorphic subgraphs to reduce computational overhead. The privacy-preserving capability of these k-anonymity methods depends heavily on the value of k. However, a larger k would inevitably lead to a lower data utility.

Approaches based on the node or edge perturbation create rules for processing nodes and edges, including edge modification, node clustering, and random walk. Node and edge approaches are achieved by inserting or removing nodes and edges. Hay *et al.* [69] proposed a low-complexity random perturbation method by randomly removing and inserting edges on the graph. Yu *et al.* [174] proposed a perturbation strategy based on local clustering, where edges were modified to mitigate the risk of privacy compromise while minimizing the loss of network structures and data.

DP was proposed to quantify the privacy loss of individuals whose data undergoes algorithmic processing [48]. Graph DP is a special case of DP. It is typically divided into node-DP and edge-DP, based on the addition and deletion of a single node or edge. In this sense, graph DP extends the node and edge perturbation approaches by quantifying their privacy loss. Day *et al.* [38] explored the graph degree distribution under node DP by designing a projection technique that incorporates aggregation and cumulative histogram to reduce the degree sensitivity. Huang *et al.* [75] developed a privacy protection algorithm using the adjacency degrees of a graph, which combined clustering and randomization. The privacy of the graph was protected by clustering, reconstructing the graph based on ordered degrees, and generating noisy nodes. Li *et al.* [102] proposed a general subspace perturbation approach for privacy preservation and distributed optimization, where noises were injected into the non-convergent subspace through dual variables such that the private data was protected. Compared with k-anonymity, DP can protect the statistics of graph data, e.g., the degree distribution and edge weights.

Only a few studies have been dedicated to preserving the privacy of graph data with latent graph structures, more explicitly, the latent graph structures [70, 101]. To limit the disclosure probability, He *et al.* [70] developed a privacy analysis framework with a noise-adding process for the average consensus algorithm. Li *et al.* [101] addressed the privacy issues using a distributed graph filtering technique, which provides each node with the ability to compute its own desired output while maintaining its privacy. However, these methods [70, 101] only considered the privacy of the data and did not consider the privacy of the latent graph structure underlying the graph data.

2.1.1 Identifying Privacy Risks in Graph Data

Privacy risks [162] in graph data are significant due to the inherent structure and interconnected nature of graphs, which can reveal sensitive information about individuals and their relationships. One primary risk is node attribute disclosure, where sensitive information associated with individual nodes can be inferred from the graph structure or neighboring nodes. This can happen even if the data appears anonymized because the unique patterns and connections in the graph can provide clues to an individual's identity or personal attributes.

Another significant risk is link inference, where the presence or absence of edges between nodes can reveal confidential relationships, such as social connections, professional collaborations, or financial transactions. This becomes particularly problematic in social networks and organizational graphs where relationships are central to the data's purpose. In addition, re-identification attacks pose a threat as attackers can potentially re-identify anonymized nodes by matching graph patterns with external data, exploiting the unique structure of the graph to uncover identities.

Neighborhood attacks further exacerbate privacy risks by allowing attackers to infer sensitive information about a node based on its neighbors' attributes and connections. Similarly, subgraph disclosure risks arise when specific patterns or subgraphs within the larger graph correspond to sensitive groups or hierarchies, making it possible to expose these structures through analysis. Aggregation risks occur when combining data from multiple sources into a single graph, which can inadvertently reveal more information than intended due to the comprehensive view it provides.

Inference attacks add another layer of complexity, as machine learning models trained on graph data can unintentionally leak sensitive information. These models can be exploited to infer private data from their outputs, posing significant privacy concerns. Understanding these various privacy risks is crucial for developing robust privacy-preserving techniques in graph learning, ensuring that sensitive information is protected while still allowing valuable insights to be derived from graph data.

2.1.2   Sensitivity of Graph Signals or Graph-Structured Data

Graph signals and graph-structured data are highly sensitive due to the detailed and interconnected information they encapsulate. Seeing as graph-structured data captures information not just about individuals themselves, but about their relationships with other participants, all of these attacks can potentially compromise the privacy of multiple participants at once. Each node in a graph can hold extensive personal or confidential information, such as attributes, transactions, or behavioral metrics. This makes the data highly sensitive, especially when representing individuals or sensitive entities.

The connections between nodes, representing relationships or interactions, add another layer of sensitivity. These edges reveal complex interdependencies, such as social networks, professional associations, or communication patterns, which can expose private details about how individuals or entities are related. Additionally, the unique structure of a graph, with its specific arrangement of nodes and edges, can act as an identifying feature. Even anonymized graphs can sometimes be re-identified by matching their structure to known graphs, posing significant privacy risks.

Dynamic graphs, which evolve over time, add another dimension to the sensitivity of graph data. Temporal changes in the graph can indicate the development or dissolution of relationships, providing insights into sensitive temporal dynamics, such as shifts in social or financial interactions. Furthermore, machine learning models trained on graph data can infer sensitive relationships or attributes, as predictive models might reveal private information based on nodes' positions and connections within the graph.

Analyzing graph data can produce aggregate insights that, while helpful, can inadvertently expose sensitive information. For instance, algorithms that detect communities within graphs can reveal tightly-knit groups corresponding to private social circles or exclusive professional networks. This can lead to unintended disclosure of sensitive group memberships or affiliations.

2.1.3   Privacy Threats in Learning Graph Structures

Learning graph structures poses significant privacy threats due to the detailed and interconnected nature of graphs, which capture complex relationships and attributes of nodes. One primary threat is node re-identification, where anonymized nodes can often be re-identified by matching unique patterns or subgraphs within the anonymized graph to known external information. This threat is particularly severe in social networks and other contexts where external data is readily available for cross-referencing, making it challenging to ensure true anonymity.

Another significant threat is link disclosure, where the presence or absence of edges (links) between nodes can reveal sensitive relationships. In a social network, for example, the existence of a link might indicate a friendship or professional relationship, while its absence might reveal a lack of association. Learning graph structures can inadvertently disclose these links, compromising the privacy of the relationships represented. Furthermore, attributes of nodes, such as demographic information, interests, or behaviors, can be inferred through their connections and the overall structure of the graph. Even if direct attributes are not disclosed, attackers can use graph learning algorithms to infer sensitive attributes based on patterns and clusters within the graph, leading to indirect information leakage.

Neighborhood attacks exploit the local neighborhood of a node to expose sensitive information about that node. For instance, if most neighbors of a node share a specific attribute, it can be inferred that the node might also possess that attribute. This type of attack leverages the homophily property of graphs, where similar nodes tend to be connected. Similarly, subgraph matching techniques can identify specific patterns within a larger graph corresponding to sensitive structures, such as organizational hierarchies or exclusive social groups, exposing the nodes involved in these structures.

Global structure inference can reveal sensitive macro-level information about the graph, such as the centrality of certain nodes, community structures, or connectivity patterns. Learning and analyzing these global structures can lead to the exposure of influential individuals, key connectors, or hidden communities, which can be sensitive in various contexts. In dynamic graphs, which evolve over time, learning the changes in the graph structure can expose temporal patterns of interactions or relationships. For example, frequent communication between nodes over time might reveal business partnerships or personal relationships, adding another layer of sensitivity.

Model inversion attacks pose another threat, where machine learning models trained on graph data are used by attackers to infer the training data. This can lead to the exposure of sensitive information about the nodes and their relationships that were used to train the model. Given these privacy threats, it is crucial to develop effective mitigation strategies to protect sensitive information within graph data.

2.2   Basics of Differential Privacy

DP [179, 173] was proposed to objectively quantify the privacy loss of individuals whose data is subjected to algorithmic processing. The DP framework and its associated techniques allow data analysts to draw conclusions about datasets while preserving individuals' privacy.

2.2.1   Definition and Principles of Differential Privacy

Differential privacy ensures that the inclusion or exclusion of a single individual's data does not significantly affect the outcome of any analysis. This means that an observer cannot determine whether any specific individual's information is part of the dataset by looking at the analysis results, thus protecting individual privacy.

Let $\mathbf{D}_1$ be a database and $\mathbf{D}_2$ be a copy of $\mathbf{D}_1$ that is different in any one tuple. $\mathbf{D}_1$ and $\mathbf{D}_2$ are neighboring databases. The definitions and properties of differential privacy are described as follows.

Definition 2.2.1 *($\epsilon - $ **DP**): A randomized algorithm $\mathcal{A}$ is differentially private if for any subsets $S \in Range(\mathcal{A})$ and all neighboring datasets $\mathbf{D}_1$ and $\mathbf{D}_2$ satisfies*

$$\Pr[\mathcal{A}(D_1) \in S] \leq e^\epsilon \cdot \Pr[\mathcal{A}(D_2) \in S]. \tag{2.1}$$

ϵ is a non-negative parameter that controls the privacy level. A smaller ϵ means better privacy but potentially less accurate results, while a larger ϵ allows more accurate results but weaker privacy guarantees. The choice of ϵ involves a trade-off between privacy and utility.

Differential privacy ensures that the output of a data analysis algorithm is almost equally likely, regardless of whether any single individual's data is included in the dataset. This principle guarantees that individual data points do not have a significant impact on the analysis results, thus maintaining privacy.

2.2.2   Differential Privacy in Graphs

Differential privacy is a robust privacy-preserving framework that can be applied to graph data to protect the privacy of individuals or entities represented within the graph. Given the complex and interconnected nature of graph structures, implementing differential privacy in graphs involves specific techniques and considerations.

In a setting of DP on graph-structured data, a graph algorithm $\mathcal{A}$ is assumed to be ϵ-node differentially private if for any two graphs $\mathcal{G}_1$ and $\mathcal{G}_2$ that differ by exactly one node and all its incident edges, and for all subsets of outputs $S \in \text{Range}(\mathcal{A})$

$$\Pr[\mathcal{A}(G_1) \in S] \leq e^{\epsilon} \cdot \Pr[\mathcal{A}(G_2) \in S]. \tag{2.2}$$

From $\mathbf{D}_1$ a neighboring dataset $\mathbf{D}_2$ is constructed by either (a) removing or adding one node and its adjacent edges (node-level DP) or (b) removing or adding one edge (edge-level DP).

Node differential privacy aims to protect the presence or absence of individual nodes in the graph, along with their incident edges. This type of privacy is particularly useful when nodes represent sensitive entities, such as individuals in a social network or patients in a healthcare network. Ensuring node differential privacy typically requires more stringent measures compared to edge differential privacy due to the potential higher sensitivity of node data.

Definition 2.2.2 (*Node-level DP*): *Under node-level DP, two graphs* $\mathcal{G}_1 = (\mathcal{V}_1, \mathcal{E}_1)$ *and* $\mathcal{G}_2 = (\mathcal{V}_2, \mathcal{E}_2)$ *are defined as neighboring if they differ in a single node and its corresponding edges (achieved through a node removal/addition)[160], ϵ-node differential privacy is therefore preserved if (2.2) holds for all events S and all pairs of neighbors $\mathcal{G}_1$, $\mathcal{G}_2$, that differ in a single node and its corresponding edges:*

$$\mathcal{V}_2 = \mathcal{V}_1 \setminus v_i \wedge \mathcal{E}_2 = \mathcal{E}_1 \setminus c, \tag{2.3}$$

where v_i is a node in $\mathcal{V}_1$ and c is the set of all edges connected to v_i.

Edge differential privacy focuses on protecting the presence or absence of individual edges in the graph. This is useful when the relationships or interactions between nodes are sensitive. Edge differential privacy is often less restrictive than node differential privacy because it deals with smaller changes to the graph's structure.

Definition 2.2.3 (*Edge-level DP*): *Under edge-level DP, two graphs $\mathcal{G}_1 = (\mathcal{V}_1, \mathcal{E}_1)$ and $\mathcal{G}_2 = (\mathcal{V}_2, \mathcal{E}_2)$ are defined as neighboring if they differ in a single edge (either through addition or through the removal of the edge), ϵ-edge differential privacy is therefore preserved if (2.2) holds for all events S and all pairs of neighbors $\mathcal{G}_1$, $\mathcal{G}_2$, that differ in a single edge, thus, $\mathcal{G}_1$ and $\mathcal{G}_2$ are neighbors if*

$$\mathcal{V}_2 = \mathcal{V}_1 \wedge \mathcal{E}_2 = \mathcal{E}_1 \setminus e_i, \tag{2.4}$$

where $e_i \in \mathcal{E}_1$.

2.2.3 Noise Addition for Signal Privacy

To achieve differential privacy, noise must be added to the algorithm output. The amount of noise required depends on the sensitivity of the function being computed. Let's assume that $\mathcal{A}$ performs a function (or query) f on the graph dataset. In the context of differential privacy in GNNs, f represents a repeated process involving the forward pass, loss computation, and gradient calculation of the GNNs. To determine the appropriate level of noise to add, it is necessary to calculate the sensitivity of f, to which the noise will be applied. Formally, the sensitivity Δf of a function f is defined as:

$$\Delta f = \max_{\mathbf{D}_1, \mathbf{D}_2 \in \mathbf{X}, \mathbf{D}_1 \simeq \mathbf{D}_2} \|f(\mathbf{D}_1) - f(\mathbf{D}_2)\|. \tag{2.5}$$

To achieve differential privacy, noise is added to the output of functions computed on the data. The Laplace mechanism is one of the primary methods used to achieve differential privacy by adding noise to the output of a function based on the Laplace distribution. This mechanism ensures that the output does not reveal sensitive information about any individual data point in the dataset.

Definition 2.2.4 *(**Laplace Mechanism**): The Laplace mechanism adds noise drawn from the Laplace distribution to the output of a function to achieve ϵ-differential privacy. For a given function f and dataset $\mathbf{D}_1$, the differentially private output is computed as:*

$$\mathcal{A}(\mathbf{D}_1) = f(\mathbf{D}_1) + Lap\left(\frac{\Delta f}{\epsilon}\right). \tag{2.6}$$

Here, $Lap\left(\frac{\Delta f}{\epsilon}\right)$ denotes the Laplace noise with scale parameter $\frac{\Delta f}{\epsilon}$.

The Gaussian mechanism is another fundamental approach used to achieve differential privacy, particularly useful when a more nuanced trade-off between privacy and utility is required. Unlike the Laplace mechanism, which adds noise from the Laplace distribution, the Gaussian mechanism adds noise drawn from the Gaussian (normal) distribution.

Definition 2.2.5 *(**Gaussian Mechanism**): The Gaussian mechanism adds noise drawn from the Gaussian distribution to the output of a function to achieve ϵ-differential privacy. For a given function f and dataset $\mathbf{D}_1$, the differentially private output is computed as:*

$$\mathcal{A}(\mathbf{D}_1) = f(\mathbf{D}_1) + \mathcal{N}(0, \sigma^2). \tag{2.7}$$

Here, $\mathcal{N}(0, \sigma^2)$ denotes the Gaussian noise with zero-mean and variance σ^2. The standard deviation σ of the Gaussian noise is chosen based on the desired privacy parameters ϵ and δ, and the sensitivity Δf. Specifically, the variance σ^2 is set to

$$\sigma = \frac{\Delta f \sqrt{2\ln(1.25/\delta)}}{\epsilon}. \tag{2.8}$$

2.3 Privacy Preserving for Graphs

Numerous privacy-preserving techniques for graph have been developed, e.g., k-anonymity [175], node and edge perturbation [69], and graph DP [48]. These methods do not apply to graph data, i.e., the three-dimensional data with latent graph structures (or, in other words, a set of concurrent time series), such as the brain signals. The methods would require the latent graphs to be extracted from the data, and separately obfuscated. The usefulness or effectiveness of graph data would degrade.

Yuan *et al.* [175] modeled a k-degree-l-diversity anonymity system and proposed an anonymization method that adds noisy nodes into the original graph to protect structural information. k measures the number of vertices with the same degrees. l indicates the distinct labels. Ding *et al.* [43] developed a privacy-preserving framework to anonymize graphs, which defined an information loss matrix for graph datasets based on a k-decomposition method. k-anonymity was performed on isomorphic subgraphs to reduce computational overhead. The privacy-preserving capability of these k-anonymity methods is heavily influenced by the magnitude of k. However, a higher k value would inevitably lead to a lower data utility.

Approaches based on the node or edge perturbation create rules for processing nodes and edges, including edge modification, node clustering, and random walk. Node and edge approaches are achieved by inserting or removing vertices and edges. Hay *et al.* [69] developed a low-complexity random perturbation method by randomly removing and inserting edges on the graph. Yu *et al.* [174] introduced a perturbation scheme derived from local clustering, where edges were modified to mitigate the risk of privacy compromise while minimizing the loss of network structures and data.

DP was proposed to quantify the privacy loss of individuals whose data undergoes algorithmic processing [48]. Graph DP is a special case of DP. It is typically divided into node-DP and edge-DP, built upon the concept of adding and removing a single node or edge. In this sense, graph DP extends the node and edge perturbation approaches by quantifying their privacy loss. Day *et al.* [38] explored the graph degree distribution under node DP by designing a projection technique that incorporates aggregation and cumulative histogram to reduce the degree sensitivity. Huang *et al.* [75] developed a privacy protection algorithm using the adjacency degrees of a graph, combining clustering and randomization. The privacy of the graph was protected by clustering, reconstructing the graph based on ordered degrees, and generating noisy nodes. Li *et al.* [102] proposed a general method for perturbing subspace to achieve privacy preservation and distributed optimization, where noises were injected into the non-convergent subspace using the dual variables to preserve the privacy information. Compared with k-anonymity, DP can protect the statistics of graph data, e.g., the degree distribution and edge weights.

Only a few studies have been dedicated to preserving the privacy of graph data with latent graph structures, more explicitly, the latent graph structures [70, 101]. To limit the disclosure probability, He *et al.* [70] developed a privacy analysis framework with a noise-adding process for the average consensus algorithm. Li *et al.* [101] addressed the privacy issues using a distributed graph filtering technique, which provides each node with the ability to compute its own desired output while maintaining its privacy. However, these methods [70, 101] only considered the privacy of graph data without taking into account the privacy of the latent graph structure underlying the graph data.

3

Existing Technologies of Graph Learning

Constructing a meaningful graph topology is essential for effectively representing, processing, analyzing, and visualizing structured data. Graph learning techniques encompass a variety of methods designed to extract meaningful patterns and insights from graph-structured data. These techniques include statistical methods like correlation networks and graphical lasso, which analyze relationships between nodes. Furthermore, learning graphs from smooth signals involves methods such as Laplacian-based factor analysis, which leverages the smoothness of signals over the graph structure. Techniques for learning from stationary and band-limited graph processes further refine graph structures based on observed data patterns.

This chapter explores existing technologies of graph learning. It covers statistical methods such as correlation networks and graphical lasso, which analyze node relationships. Additionally, it discusses techniques for learning graphs from smooth signals, including Laplacian-based factor analysis and signal smoothness with edge sparsity, as well as methods for learning from stationary and band-limited graph processes to refine and understand graph structures.

3.1 Statistical Methods

Statistical methods in graph learning are foundational approaches that leverage statistical principles to analyze and infer relationships within graph-structured data. These methods provide a robust framework for understanding complex networks by modeling dependencies and interactions between nodes and edges.

3.1.1 Correlation Networks

The most commonly used linear measure of similarity between nodal random variables x_i and x_j is the Pearson correlation coefficient, which is defined as

$$\rho_{x_i, x_j} = \frac{\mathrm{Cov}(x_i, x_j)}{\sigma_{x_i} \sigma_{x_j}} = \frac{\mathrm{Cov}(x_i, x_j)}{\sqrt{\mathrm{Var}(x_i), \mathrm{Var}(x_j)}}, \tag{3.1}$$

DOI: 10.1201/9781003516613-3

where $\mathrm{Cov}(x_i, x_j)$ is the covariance of x_i and x_j of the random graph signal $\mathbf{x} = [x_1, \cdots, x_I]$, and σ_{x_i} and σ_{x_j} are the standard deviations of x_i and x_j, respectively. Given this choice, it is natural to define the correlation network $\mathcal{G} = (\mathcal{V}, \mathcal{E}, \mathbf{W})$ with vertices $\mathcal{V} = \{1, \cdots, N\}$ and edge set $\mathcal{E} = \{(i,j) \in \mathcal{V} \times \mathcal{V} : \rho_{ij} \neq 0\}$. The definition of the weights allows for some flexibility. To directly represent the correlation strength between x_i and x_j, one can define $W_{ij} = |\rho_{ij}|$ or its un-normalized variant $W_{ij} = |\mathrm{Cov}(x_i, x_j)|$. Using the covariance directly as the weight preserves the scale of the data, which can be useful in certain applications where the magnitude of covariance is significant. By defining the weights in these ways, the resulting correlation network effectively represents the strength and nature of the relationships between the variables, facilitating further analysis and learning tasks on the graph.

Correlation networks are limited in their ability to capture only linear and symmetric pairwise dependencies among vertex-indexed random variables. Crucially, the observed correlations might arise from latent network effects rather than direct interactions between vertices. For example, a presumed regulatory interaction between genes i and j, inferred from their highly correlated microarray expression profiles, could actually be the result of a third hidden gene k that regulates both i and j. Therefore, if the goal is to construct a graph that accurately reflects direct influences between pairwise signal elements, correlation networks might not be the most suitable choice. [116] proposed to resolve such a confounding by instead considering partial correlations

$$\rho_{x_i, x_j} = \frac{\mathrm{Cov}(x_i, x_j | \mathcal{V} \setminus ij)}{\sigma_{x_i} \sigma_{x_j}} = \frac{\mathrm{Cov}(x_i, x_j)}{\sqrt{\mathrm{Var}(x_i | \mathcal{V} \setminus ij), \mathrm{Var}(x_j | \mathcal{V} \setminus ij)}}, \qquad (3.2)$$

where $\mathcal{V} \setminus ij$ symbolically denotes the collection of all $N - 2$ random variables $\{x_k\}$ after excluding those indexed by nodes i and j.

3.1.2 Graphical Lasso

Graphical Lasso [60] is a statistical technique used in graph learning to estimate sparse inverse covariance matrices, representing the conditional dependencies between variables in a graph. This method is precious in scenarios where the number of variables is large, such as in high-dimensional datasets, relative to the number of observations. By promoting sparsity, graphical Lasso helps uncover significant relationships while ignoring spurious ones, leading to more straightforward and more interpretable graph structures.

The core idea behind Graphical Lasso is to apply ℓ_1-regularization to the elements of the inverse covariance matrix, encouraging many of the entries to be zero. This results in a sparse precision matrix where non-zero entries indicate direct conditional dependencies between variables. The optimization problem for Graphical Lasso can be defined as:

$$\hat{\Theta} = \arg\min_{\Theta \succ 0} \left(\text{tr}(\Sigma\Theta) - \log\det\Theta + \lambda\|\Theta\|_1 \right), \qquad (3.3)$$

where Θ is the precision matrix to be estimated; Σ is the empirical covariance matrix of the observed data; λ is a regularization parameter that controls the sparsity level of Θ; $\text{tr}(\Sigma\Theta)$ is the trace of the product of the empirical covariance matrix and the precision matrix; $\log\det\Theta$ is the log-determinant of the precision matrix, ensuring that Θ is positive definite; $\|\Theta\|_1$ is the ℓ_1-norm of Θ, summing the absolute values of its entries to promote sparsity.

Graphical Lasso enhances interpretability by producing a sparse precision matrix, simplifying the graph structure and making it easier to identify direct relationships between variables. This is especially valuable in fields like genomics [157] and neuroscience [77, 171], where understanding direct interactions is crucial. On the other hand, graphical Lasso is scalable due to the development of efficient algorithms that allow its application to large-scale datasets, making it a practical tool for high-dimensional data analysis. Furthermore, the flexibility of the graphical Lasso enables it to be adapted to various types of data and research questions, from identifying gene regulatory networks in biology [187] to modeling financial asset dependencies in economics [8], making it a versatile technique in graph learning.

3.2 Learning Graphs from Observations of Smooth Signals

According to Section 1.2.4 in Chapter 1, the signal is smooth if the signal values associated with the two end vertices of edges with large weights in the graph tend to be similar. This section presents an algorithm for learning a valid graph Laplacian operator from data samples, such that the graph signal representation is smooth and consistent with the Gaussian prior on the latent variables.

3.2.1 Laplacian-Based Factor Analysis Model

The factor analysis model [13] is a generic linear statistical model that tries to explain observations of a given dimension with a potentially smaller number of unobserved latent variables. It was proposed in [44] to estimate graph Laplacians, aiming for the input graph signals to be smooth across the derived topologies. Specifically, suppose $\mathbf{L} = \mathbf{U}\mathbf{\Lambda}\mathbf{U}^T$ is the eigendecomposition of the combinational Laplacian associated with an unknown associated, undirected graph $\mathcal{G}(\mathcal{V}, \mathcal{E}, \mathbf{W})$ with $N = |\mathcal{V}|$ vertices. The observed graph signal $\mathbf{x} \in \mathbb{R}^N$ is assumed to have zero means for simplicity and follows graph-dependent factor analysis model [44]:

$$\mathbf{x} = \mathbf{U}\chi + \epsilon, \tag{3.4}$$

where the factors are given by the Laplacian eigenvectors $\mathbf{U}$, $\chi \in \mathbb{R}^N$ represents latent variables or factor loadings, and $\epsilon \sim \text{Norm}(\mathbf{0}, \sigma^2 \mathbf{I})$ is an isotropic error term. The Laplacian eigenvectors provide a spectral embedding of the graph vertices, which is valuable for advanced network analysis tasks like data visualization, clustering, and community detection. The representation matrix creates an initial connection between the signal model and the graph topology. The second connection arises from the prior distribution of the adopted latent variables, i.e., $\chi \sim \text{Norm}(\mathbf{0}, \mathbf{\Lambda}^{\dagger})$, where $\dagger$ is pseudoinverse and thus the precision matrix is defined as the eigenvalue matrix $\mathbf{\Lambda}$ of the Laplacian. The GFT interpretation of (3.4) implies that the prior on χ promotes low-pass bandlimited $\mathbf{x}$. Specifically, the mapping $\mathbf{\Lambda} \to \mathbf{\Lambda}^{\dagger}$ converts the large Laplacian eigenvalues (corresponding to high frequencies) into low-power factor loadings in χ. Conversely, small eigenvalues related to low frequencies are mapped to high-power factor loadings. This reflects the model's imposition of smoothness before $\mathbf{x}$.

Given the observed signal $\mathbf{x}$, the maximum a posteriori (MAP) estimator of the latent variables is defined as (σ^2 is later considered known and incorporated into the parameter $\alpha > 0$):

$$\chi_{\text{MAP}} = \arg \min_{\chi} \left\{ \|\mathbf{x} - \mathbf{U}\chi\|^2 + \alpha \chi^T \mathbf{\Lambda} \chi \right\}, \tag{3.5}$$

which is naturally dependent on the unknown eigenvectors and eigenvalues of the Laplacian. With $\mathbf{y} := \mathbf{U}\chi$ representing the predicted graph signal (or error-free version of $\mathbf{x}$), it follows that

$$\chi^T \mathbf{\Lambda} \chi = \mathbf{y}^T \mathbf{U} \mathbf{\Lambda} \mathbf{U}^T \mathbf{y} = \mathbf{y}^T \mathbf{L} \mathbf{y} = \text{TV}(\mathbf{y}). \tag{3.6}$$

Consequently, the MAP estimator (3.5) can be viewed as a Laplacian-based TV denoiser for $\mathbf{x}$, effectively imposing a smoothness prior on the recovered signal $\mathbf{y} = \mathbf{U}\chi$. Furthermore, (3.5) can be interpreted as a kernel ridge-regression estimator with the (unknown) Laplacian kernel $\mathbf{K} := \mathbf{L}^{\dagger}$.

Building upon the formulation described in (3.5), with the incorporation of graph topology as a variable in the optimization process, the objective is to simultaneously optimize the graph Laplacian $\mathbf{L}$ and achieve a denoised representation $\mathbf{y} = \mathbf{U}\chi$ of $\mathbf{x}$. This approach aims to solve

$$\min_{\mathbf{L}, \mathbf{y}} \left\{ \|\mathbf{x} - \mathbf{y}\|^2 + \alpha \mathbf{y}^T \mathbf{L} \mathbf{y} \right\}. \tag{3.7}$$

The objective function described in (3.7) serves two main purposes: 1) ensuring data fidelity by penalizing the quadratic loss between $\mathbf{y}$ and the observed $\mathbf{x}$, and 2) promoting smoothness on the graph structure using TV regularization.

When considering multiple independent observations $\mathcal{X} := \{\mathbf{x}_p\}_{p=1}^P$ organized in the matrix $\mathbf{X} = [\mathbf{x}_1, \ldots, \mathbf{x}_P] \in \mathbb{R}^{N \times P}$, the approach outlined in [44] aims to solve

$$\min_{\mathbf{L},\mathbf{Y}} \left\{ \|\mathbf{X} - \mathbf{Y}\|_F^2 + \alpha \mathrm{trace}(\mathbf{Y}^T \mathbf{L} \mathbf{Y}) + \frac{\beta}{2} \|\mathbf{L}\|_F^2 \right\}$$

$$\text{s. t.} \quad \mathrm{trace}(\mathbf{L}) = N, \quad \mathbf{L}\mathbf{1} = 0, \quad L_{ij} = L_{ji} \leq 0, \quad i \neq j. \qquad (3.8)$$

The optimization framework includes constraints on $\mathbf{L}$ to ensure it qualifies as a valid combinatorial Laplacian. Specifically, enforcing $\mathrm{trace}(\mathbf{L}) = N$ prevents the trivial all-zero solution and effectively fixes the ℓ_1-norm of $\mathbf{L}$. To regulate the sparsity of the resultant graph, a Frobenius-norm penalty is incorporated into the objective function of (3.7) to reduce its edge weights. The balance between data fidelity, smoothness, and sparsity is controlled through positive regularization parameters α and β.

Although (3.8) is not jointly convex in $\mathbf{L}$ and $\mathbf{Y}$, it exhibits bi-convexity. This property ensures that when $\mathbf{L}$ is fixed, the problem with respect to $\mathbf{Y}$ is convex, and vice versa. Therefore, the algorithmic approach outlined in [44] employs alternating minimization, a technique that iteratively minimizes either $\mathbf{L}$ or $\mathbf{Y}$ while holding the other fixed, converging to a stationary point of (3.8). When $\mathbf{L}$ is fixed, (3.8) simplifies to a quadratic program (QP) subject to linear constraints, which can be efficiently solved using interior point methods. When $\mathbf{L}$ is fixed, the resulting problem is a matrix-valued counterpart of (3.7). The closed-form solution is given by $\mathbf{Y} = (\mathbf{I} + \alpha\mathbf{L})^{(-1)}\mathbf{X}$, which represents a low-pass, graph filter-based smoother of the signals in $\mathbf{X}$.

3.2.2 Signal Smoothness with Edge Sparsity

A different method for addressing the challenge of learning graphs with a smoothness prior was introduced in [86]. Consider the data matrix $\mathbf{X} = [\mathbf{x}_1, \ldots, \mathbf{x}_P] \in \mathbb{R}^{N \times P}$, with $\mathbf{x}_i^T \in \mathbb{R}^{1 \times P}$ representing the ith row, which gathers the P measurements at vertex i. The main concept in [86] is to demonstrate the connection between smoothness and sparsity, highlighted through the identity. The objective function is defined as

$$\sum_{p=1}^{P} \mathrm{TV}(\mathbf{x}_p) = \mathrm{trace}(\mathbf{X}^T \mathbf{L} \mathbf{X}) = \frac{1}{2} \|\mathbf{W} \circ \mathbf{Z}\|_1, \qquad (3.9)$$

where the Euclidean-distance matrix $\mathbf{Z} \in \mathbb{R}_+^{N \times N}$ has entries $Z_{ij} := \|\mathbf{x}_i - \mathbf{x}_j\|_2^2$, $i, j \in \mathcal{V}$. The idea is that, if the distances in $\mathbf{Z}$ are derived from a smooth manifold, the resulting graph will have a sparse edge set, favoring edges (i, j) with smaller distances Zij. Identity (3.9) provides an effective way to parameterize graph-learning formulations under smoothness constraints because adjacency matrices can be described using simpler, entry-wise decoupled constraints compared to their Laplacian counterparts. It also shows that when a smoothness penalty is applied in the search criterion for $\mathcal{G}$, adding additional sparsity-inducing regularization becomes redundant.

Given these considerations, a general-purpose model for learning graphs is advocated in [86], i.e.,

$$\min_{\mathbf{W}} \left\{ \|\mathbf{W} \circ \mathbf{Z}\|_1 - \alpha \mathbf{1}^T \log(\mathbf{W1}) + \frac{\beta}{2} \|\mathbf{W}\|_F^2 \right\}$$

$$\text{s. t.} \quad \text{diag}(\mathbf{W}) = 0, \ W_{ij} = W_{ji} \geq 0, \ i \neq j, \tag{3.10}$$

where α and β are tunable regularization parameters. Unlike [44], the logarithmic barrier on the vector $\mathbf{W1}$ of nodal degrees enforces each vertex to have at least one incident edge. The Frobenius-norm regularization on the adjacency matrix $\mathbf{W}$ controls the graph's edge sparsity pattern by penalizing larger edge weights. Overall, this combination forces degrees to be positive but does not prevent most individual edge weights from becoming zero. The sparsest graph is obtained when $\beta = 0$, and edges form preferentially between nodes with smaller Z_{ij}, similar to a 1-nearest neighbor graph. The convex optimization problem (3.4) can be efficiently solved with a complexity of $\mathcal{O}(N^2)$ per iteration by utilizing provably convergent primal-dual solvers that support parallelization.

3.3   Learning Graphs from Observations of Stationary Graph Process

Many tools in graph signal processing leverage the structure of the underlying graph, such as the Laplacian matrix, to process signals on the graph. Consequently, if a graph is not available, these tools become unusable. Researchers have developed methods to address this by inferring graph topology from observations of signals at its vertices. Given the ill-posed nature of the problem, these methods typically assume properties such as signal smoothness on the graph or sparsity priors. Pasdeloup *et al.* [131] proposed a characterization of the space of valid graphs that can account for stationary signals. The main focus of this approach is on identifying graphs that explain the structure of a random signal.

As in previous sections, the goal is to infer the symmetric graph shift $\mathbf{B} = \mathbf{U}\mathbf{\Lambda}\mathbf{U}^T$ associated with the unknown underlying graph $\mathcal{G}$ from the graph signal $\mathcal{X} := \{\mathbf{x}_p\}_{p=1}^P$. It is assumed that each $\mathbf{x}_p$ originates from a network diffusion process influenced by $\mathbf{B}$. Formally, the approach considers a random network process $\mathbf{x} = \sum_{l=0}^{L-1} h_l \mathbf{B}^l \mathbf{w} = \mathbf{Hw}$ driven by a zero-mean input $\mathbf{w}$. It is currently assumed that $\mathbf{w}$ is white, with $\mathbb{E}[\mathbf{w}\mathbf{w}^T] = \mathbf{I}$. The primary objective is to deduce the direct relationships described by $\mathbf{B}$ from the set $\mathcal{X}$ comprising P independent samples of the random signal $\mathbf{x}$. This setup assumes no prior knowledge of the filter degree $L - 1$, the coefficients $\mathbf{h}$, or the specific realizations of the inputs $\{\mathbf{w}p\}_{p=1}^{P}$.

The described problem is highly underdetermined and non-convex. It is underdetermined because each observation $\mathbf{x}_p$ introduces as many unknowns in the input $\mathbf{w}_p$, along with the unknown filter coefficients $\mathbf{h}$ and the shift $\mathbf{B}$, which is the primary focus. The nonconvexity arises because the observations are dependent on the product of these unknowns and particularly on the first $L - 1$ powers of $\mathbf{B}$. To tackle the underdeterminacy, the approach relies on the statistical characteristics of the input process $\mathbf{w}$ and imposed regularity on the graph to be reconstructed, such as edge sparsity or minimum-energy weights. To handle the nonconvexity, the overall inference task is divided by initially estimating the eigenvectors of $\mathbf{B}$, which are invariant to any power of $\mathbf{B}$ and subsequently its eigenvalues. This leads to a two-phase procedure: 1) using the observation model to infer $\mathbf{U}$ from the signals $\mathcal{X}$, and 2) combining $\mathbf{U}$ with prior information about $\mathcal{G}$ and applying feasibility constraints on $\mathbf{B}$ to derive the optimal eigenvalues.

3.3.1 Inferring the Eigenvectors

Based on the described model, the covariance matrix $\mathbf{\Sigma_x}$ of the signal $\mathbf{x}$ is given by:

$$\mathbf{\Sigma}_x = \mathbb{E}[\mathbf{H}\mathbf{w}(\mathbf{H}\mathbf{w})^T] = \mathbf{H}^2 = \mathbf{U}\left(\sum_{l=0}^{L-1} h_l \Lambda^l\right)^2 \mathbf{U}^T \qquad (3.11)$$

$$= h_0 \mathbf{I} + 2h_0 h_1 \mathbf{B} + (2h_0 h_2 + h_1^2)\mathbf{B}^2 + \cdots,$$

where $\mathbf{\Sigma}_w = \mathbf{I}$ and $\mathbf{H} = \mathbf{H}^T$ because $\mathbf{B}$ is assumed to be symmetric, it becomes evident from (3.11) that the eigenvectors (i.e., the GFT basis) of the shift $\mathbf{B}$ and the covariance $\mathbf{\Sigma}_x$ are identical. This indicates that the primary difference between $\mathbf{\Sigma}_x$, encompassing indirect relationships between signal elements, and $\mathbf{B}$, which includes only direct relationships, lies in their eigenvalues. Although the diffusion in $\mathbf{H}$ affects the eigenvalues of $\mathbf{B}$ according to the filter's frequency response, the eigenvectors $\mathbf{U}$ remain consistent in $\mathbf{\Sigma}_x$, reflecting the original spectrum. Thus, accessing $\mathbf{\Sigma}_x$ allows us to obtain $\mathbf{U}$ by performing an eigendecomposition of the covariance. However, obtaining an exact $\mathbf{\Sigma}_x$ from a finite set of signals $\mathbf{X}$ is generally infeasible. In practice, the covariance is estimated, often using the sample covariance $\hat{\mathbf{\Sigma}}_x$, resulting in a noisy version of the eigenvectors $\hat{\mathbf{U}}$. The robustness of this two-step process to the noise in $\hat{\mathbf{U}}$ is examined in Section 3.3.2.

The simultaneous diagonalizability of $\mathbf{B}$ and $\mathbf{\Sigma}_x$ implies that $\mathbf{x}$ is a stationary process on the unknown graph-shift operator $\mathbf{B}$. Consequently, the graph inference problem can be restated as finding a shift on which the observed signals are stationary.

3.3.2   Inferring the Eigenvalues

From the previous discussion, it is evident that any $\mathbf{B}$ sharing eigenvectors with Σ_x can account for the observations, implying the existence of filter co-efficients $\mathbf{h}$ that produce $\mathbf{x}$ through a diffusion process on $\mathbf{B}$. Notably, the covariance matrix Σ_x itself can generate $\mathbf{x}$ through a diffusion process, as can the precision matrix Σ_x^{-1} (representing partial correlations under Gaussian assumptions). To resolve this ambiguity, which involves selecting the eigenvalues of $\mathbf{B}$, it is assumed that the shift of interest is optimal in some sense [142]. The approach is to identify the shift operator $\mathbf{B}$ that 1) is optimal concerning criteria $f(\mathbf{B})$; 2) belongs to a convex set $\mathcal{S}$ defining the desired type of shift operator (e.g., adjacency matrix $\mathbf{W}$ or Laplacian matrix $\mathbf{L}$); and 3) has the prescribed $\mathbf{U}$ as eigenvectors. Formally, this can be expressed as:

$$\min_{\mathbf{B} \in \mathcal{S}, \Lambda} f(\mathbf{B}), \quad \text{s.to } \mathbf{B} = \mathbf{U}\Lambda\mathbf{U}^T, \tag{3.12}$$

which is a convex optimization problem provided $f(\mathbf{B})$ is convex.

The selection of $f(\mathbf{B})$ allows for the integration of the desired graph's physical characteristics into the formulation while maintaining consistency with the spectral basis $\mathbf{U}$. For example, the matrix (pseudo) norm $f(\mathbf{B}) = \|\mathbf{B}\|_0$, which counts the number of nonzero entries in $\mathbf{B}$, can be utilized to minimize the number of edges. Meanwhile, $f(\mathbf{B}) = \|\mathbf{B}\|_1$ serves as a convex proxy for the edge cardinality function. Alternatively, adopting the Frobenius norm $f(\mathbf{B}) = \|\mathbf{B}\|_F$ minimizes the energy of the graph's edges. Additionally, $f(\mathbf{B}) = \|\mathbf{B}_\infty\|$ yields shifts $\mathbf{B}$ associated with graphs characterized by uniformly low-edge weights, which can be significant when identifying graphs under capacity constraints.

3.4   Learning Graphs from Observation of Band-limited Graph Signal Process

A fundamental aspect of the spectral representation of signals on a graph is its dependence on its topology. Specifically, for undirected graphs, the GFT is defined as the projection of the observed signal onto the space spanned by the eigenvectors of the graph Laplacian matrix. Consequently, a signal associated with different graphs will generally produce different spectra. Building on this concept, the core idea is to associate a graph topology with the observed signal to make the signal band-limited over the inferred graph or, equivalently, to make its spectral representation sparse. Enforcing this band-limited property is crucial for leveraging the full range of theories regarding sampling signals defined over graphs. A few existing studies [80, 139] have attempted to learn the latent graphs of bandlimited signals.

The goal is to learn the orthonormal transform matrix $\mathbf{U}$, the sparse matrix $\mathbf{S}$ of the signals GFT, and the underlying graph topology, captured by the Laplacian matrix $\mathbf{L}$ that admits the columns of $\mathbf{U}$ as its eigenvectors [140]. The objective function is defined as [140]:

$$\min_{\mathbf{L}, \mathbf{U} \in \mathbb{R}^{N \times N}, \mathbf{S} \in \mathbb{R}^{N \times M}} \|\mathbf{X} - \mathbf{US}\|_F^2 + f(\mathbf{L}, \mathbf{X}, \mathbf{S}) \tag{3.13a}$$

$$\text{s.t.} \quad \mathbf{UU}^T = \mathbf{I}, \mathbf{u}_1 = b\mathbf{1}, \tag{3.13b}$$

$$\mathbf{L} = \mathbf{U\Lambda U}^T, \mathbf{L} \in \mathbb{L}, \tag{3.13c}$$

$$\mathbf{\Lambda} \succeq \mathbf{0}, \Lambda_{ij} = \Lambda_{ji}, \forall i \neq j, \tag{3.13d}$$

$$\mathbf{S} \in \mathcal{B}_K. \tag{3.13e}$$

As described in Section 1.2.4, the GFT coefficient matrix $\mathbf{S}$ is block-sparse, containing multiple rows of zeros. $\mathbf{S} \in \mathcal{S}_K$, where K specifies the bandwidth of the frequency-domain representation of the observed band-limited graph signal $\mathbf{X}$. $\mathcal{B}_K$ collects all K-block sparse matrices [23]:

$$\mathcal{S}_K \triangleq \{\mathbf{S} = [\mathbf{s}_1, \cdots, \mathbf{s}_M] \in \mathbb{R}^{N \times M}, \mathbf{S}(i,:) = \mathbf{0}, i \notin \mathcal{K} \subseteq \mathcal{V}, K = |\mathcal{K}|\}, \tag{3.14}$$

where $\mathbf{S}(i,:)$ is the i-th row of $\mathbf{S}$, and $\mathcal{K} \in \mathcal{V}$.

In problem (3.13), the objective function comprises the sum of the data fitting error and a penalty function $f(\mathbf{L}, \mathbf{X}, \mathbf{S})$, which guides the graph topology to mirror the desired characteristics of the observed graph signals. The constraints in (3.13) have the following implications: a) ensures that $\mathbf{U}$ is unitary and includes a vector proportional to the vector of all ones, with $b = 1/\sqrt{N}$; b) restricts $\mathbf{L}$ to be a valid Laplacian matrix, having the columns of $\mathbf{U}$ as its eigenvectors and the diagonal entries of $\mathbf{\Lambda}$ as its eigenvalues, with the trace constraint, $p > 0$, preventing the trivial solution; c) requires $\mathbf{\Lambda}$ to be diagonal and positive semidefinite; and d) enforces the K-block sparsity of the GFT coefficient matrix $\mathbf{S}$. Unfortunately, problem (3.13) is non-convex in both its objective function and constraint set. A sub-optimal strategy is proposed to find an efficient numerical solution, which decouples (3.13) into two more straightforward optimization problems. This strategy involves two main steps: i) given the observation matrix $\mathbf{X}$, jointly learn the orthonormal transform matrix $\mathbf{U}$ and the sparse matrix $\mathbf{S}$ to minimize the fitting error $\|\mathbf{X} - \mathbf{US}\|^2$, subject to a block-sparsity constraint; and ii) given $\mathbf{U}$, infer the graph Laplacian matrix $\mathbf{L}$ that admits the columns of $\mathbf{U}$ as its eigenvectors.

3.4.1 Learning GFT Basis and Sparse Representation

The initial objective is to jointly learn the pair of matrices $(\mathbf{U}, \mathbf{S})$, up to a rotation matrix. Notably, recovering $\mathbf{S}$ from $\mathbf{Y}$ is straightforward if $\mathbf{U}$ is known, since $\mathbf{U}$ is invertible. However, since $\mathbf{U}$ is unknown in our case, recovering both $\mathbf{S}$ and $\mathbf{U}$ presents a challenging task, prone to identifiability issues. This problem can be solved by employing a method conceptually similar to sparsifying transform learning. The difference is that it enforces K-block-sparsity and requires one of the eigenvectors to be a constant vector. Specifically, by setting $\mathbf{u}_1 = \mathbf{b}_1$, all unnormalized Laplacian matrices inherently admit a constant vector as an eigenvector. Given that $\mathbf{U}$ is a unitary matrix, the fitting error can equivalently expressed as $\|\mathbf{Y} - \mathbf{US}\|_F^2 = \|\mathbf{U}^T\mathbf{Y} - \mathbf{S}\|_F^2$.

The goal is now to find the block-sparse columns $\{\mathbf{s}_i\}_{i=1}^M$ and the orthonormal vectors $\{\mathbf{u}_i\}_{i=1}^N$ that minimize this fitting error. Starting from the original problem $\mathbf{P}$, it can be reformulated as

$$\min_{\mathbf{L},\mathbf{U}\in\mathbb{R}^{N\times N},\mathbf{S}\in\mathbb{R}^{N\times M}} \|\mathbf{X} - \mathbf{US}\|_F^2 + f(\mathbf{L},\mathbf{X},\mathbf{S}) \tag{3.15a}$$

$$\text{s.t.} \quad \mathbf{UU}^T = \mathbf{I}, \mathbf{u}_1 = b1, \tag{3.15b}$$

$$\mathbf{S} \in \mathcal{B}_K. \tag{3.15c}$$

Although the objective function is convex, problem (3.15) remains non-convex due to the orthonormality and sparsity constraints. Drawing inspiration from [133], an algorithm is proposed to solve (3.15) by alternating between solving for the block-sparse matrix $\mathbf{S}$ and finding the orthonormal transform $\mathbf{U}$. This approach addresses the non-convex problem (3.15) by iteratively minimizing the objective function with respect to $\mathbf{S}$ and $\mathbf{U}$ at each step k, as follows:

$$\hat{\mathbf{S}}^{(k)} \triangleq \arg\min_{\mathbf{S}\in\mathbb{R}^{N\times M}} \|(\hat{\mathbf{U}}^{(k-1)})^T\mathbf{X} - \mathbf{S}\|_F^2$$
$$\text{s.t.} \quad \mathbf{S} \in \mathcal{B}_K, \tag{3.16}$$

$$\hat{\mathbf{U}}^{(k)} \triangleq \arg\min_{\mathbf{U}\in\mathbb{R}^{N\times N}} \|\mathbf{U}^T\mathbf{X} - \hat{\mathbf{S}}^{(k)}\|_F^2$$
$$\text{s.t.} \quad \mathbf{UU}^T = \mathbf{I}, \mathbf{u}_1 = b1, \tag{3.17}$$

where the superscript $^{(k)}$ denotes the step k of the algorithm. The algorithm iterates until a termination criterion is met, within a prescribed accuracy.

3.4.2   Learning Graph Topology

The second step is to recover the Laplacian matrix $\mathbf{L}$ from the estimated sparsifying transform $\hat{\mathbf{U}}$. Assuming that the observed graph signals are $\mathcal{K}$-bandlimited, with $|\mathcal{K}| = K$, it follows that the K columns of $\hat{U}_\mathcal{K}$ about the observed signal subspace are known. Given this scenario, it has

$$\mathbf{L}\hat{\mathbf{U}}_\mathcal{K} = \hat{\mathbf{U}}_\mathcal{K}\mathbf{C}_\mathcal{K}, \tag{3.18}$$

where $\mathbf{C}_\mathcal{K} \triangleq \Phi^{-1}\mathbf{\Lambda}_\mathcal{K}\Phi \succeq \mathbf{0}$, with Φ denoting a unitary $K \times K$ matrix and $\mathbf{\Lambda}_\mathcal{K}$ being the diagonal matrix obtained by selecting the set $\mathcal{K}$ of diagonal entries of $\mathbf{\Lambda}$.

Based on (3.13), after estimating $\hat{\mathbf{U}}_{\mathcal{K}}$ for the Laplacian eigenvectors and $\hat{\mathbf{S}}$ for the sparse representation, the graph learning problem can be expressed as follows:

$$\min_{\mathbf{L}\in\mathbb{R}^{N\times N},\mathbf{C}\in\mathbb{R}^{K\times K}} f(\mathbf{L},\mathbf{X},\hat{\mathbf{S}}) \tag{3.19}$$

subject to

$$\begin{aligned} \mathbf{L} &\in \mathbb{L}, \quad \operatorname{tr}(\mathbf{L}) = p \\ \mathbf{L}\hat{\mathbf{U}}_{\mathcal{K}} &= \hat{\mathbf{U}}_{\mathcal{K}}\mathbf{C}_{\mathcal{K}}, \quad \mathbf{C}_{\mathcal{K}} \succeq \mathbf{0}, \end{aligned} \tag{3.20}$$

where $\mathbb{L}$ is defined as in (3.13c). Various objective functions $f(\mathbf{L},\mathbf{Y},\hat{\mathbf{S}})$ have been suggested in the literature. Examples include: i) $f(L) = \|\mathbf{L}\|_0$, as presented in [142], and ii) $f(\mathbf{L},\mathbf{Y}) = \operatorname{tr}(\mathbf{Y}\mathbf{L}\mathbf{Y}^T) + \mu\|\mathbf{L}\|_F$, as discussed in [44]. In the first scenario, the aim was to identify the sparsest graph, using the eigenvectors of the shift matrix derived from principal component analysis (PCA) of the sample covariance of graph signals spreading across the network. In the second scenario, the objective was to minimize a combination of the l_2-norm total variation of the observed signal $\mathbf{Y}$, which evaluates signal smoothness, and a Frobenius norm term added to regulate (via the coefficient $\mu \geq 0$) the distribution of the off-diagonal entries of $\mathbf{L}$ [44]. It is important to note that the Laplacian trace constraint implies that high values of μ penalize large degrees, ultimately resulting in dense graphs with uniform degrees across nodes.

[140] introduced a different approach to reconstructing the graph topology, referred to as the Estimated-Signal-Aided Graph Learning (ESA-GL) algorithm. This method leverages the estimated GFT coefficients matrix $\hat{\mathbf{S}}$ obtained from Section 3.4.1. Start by noting that:

$$\operatorname{tr}(\mathbf{Y}^T\mathbf{L}\mathbf{Y}) = \operatorname{tr}(\mathbf{S}_{\mathcal{K}}^T\Lambda_{\mathcal{K}}\mathbf{S}_{\mathcal{K}}) = \operatorname{tr}(\hat{\mathbf{S}}_{\mathcal{K}}^T\mathbf{C}_{\mathcal{K}}\hat{\mathbf{S}}_{\mathcal{K}}), \tag{3.21}$$

where $\mathbf{S}_{\mathcal{K}}$ contains the rows with index in the support set of the graph signals, whereas $\hat{\mathbf{S}}_{\mathcal{K}} = \Phi^T\mathbf{S}_{\mathcal{K}}$ is the matrix estimated 3.4.1. Then, $f(\mathbf{L},\mathbf{Y},\hat{\mathbf{S}})$ is reformulated as

$$\min_{\mathbf{L}\in\mathbb{R}^{N\times N},\mathbf{C}_{\mathcal{K}}\in\mathbb{R}^{K\times K}} f_2(\mathbf{L},\hat{\mathbf{S}}_{\mathcal{K}},\mu) \triangleq \operatorname{tr}(\hat{\mathbf{S}}_{\mathcal{K}}^T\mathbf{C}_{\mathcal{K}}\hat{\mathbf{S}}_{\mathcal{K}}) + \mu\|\mathbf{L}\|_F \tag{3.22}$$

$$\text{s.t.} \quad (\mathbf{L},\mathbf{C}_{\mathcal{K}}) \in \mathcal{X}(\hat{\mathbf{U}}_{\mathcal{K}}). \tag{3.23}$$

In this instance, the resulting problem is also convex, allowing for efficient resolution.

4

Graph Extraction and Topology Learning of Band-Limited Signals

Graph learning provides an effective means to uncover the topological structure of band-limited signals, which is non-trivial though due to non-convex problem formulation and the required joint estimation of the graph Laplacian and the typically band-limited frequency-domain signals.

This chapter presents a new graph learning technique to learn the graph Laplacian of the band-limited signals. The key idea is that it shows the frequency-domain signals depend deterministically on the graph Laplacian. The joint estimation can be transferred to only concern the graph Laplacian. Alternating optimization (AO) is applied to learn the eigenvectors and eigenvalues of the graph Laplacian in an iterative manner. It reveals that the feasible solution for the eigenvectors is on a Stiefel manifold and can be efficiently solved using Stiefel manifold dual gradient descent. The eigenvalues are obtained using the alternating direction method of multipliers (ADMM). Experiments indicate that the new method can infer the graph topology of brain signals with substantially better accuracy than the prior art.

4.1 Introduction

The graph topology captures the instinct relationships of unstructured data encoded on different entries of the graph, and graph learning is an efficient technique used to uncover the latent graph topologies of data [45]. Classical methods for graph topology inference, such as graph lasso [60] and covariance selection [40], focused on estimating the covariance matrices of graph signals. More recent graph learning techniques, like those in [44, 86, 62, 135, 28, 142, 131], have emphasized enforcing smoothness on graph signals before topology inference. Typically, it has been assumed that the frequency-domain representations of graph signals have unlimited bandwidths for mathematical simplicity.

DOI: 10.1201/9781003516613-4

Latent graph topologies allow for the transformation of captured data into the frequency domain, facilitating effective data processing. This is crucial for tasks such as recovering missing data within a network [21] or verifying data authenticity [58]. To ensure generality, it is practical to consider band-limited graph signals, where the frequency-domain representation has finite bandwidth. This band-limited nature is observed in practice, such as in fMRI data from brain networks [72]. Smooth graph signals with unlimited frequency-domain bandwidths [44] can be seen as a special case of band-limited graph signals.

Existing graph learning methods struggle to accurately and efficiently infer the graph topology, such as the graph Laplacian [177], of band-limited graph signals due to the challenges in jointly estimating both the frequency-domain representation and the GFT basis needed to convert captured data to the frequency domain.

4.2 System Model

A network can be divided into a set of N nodes. It can be assumed that the i-th time series corresponding to these N nodes is represented by a vector, denoted by $\mathbf{y}_I \in \mathbb{R}^{N \times 1}$. Let M be the number of signals. Let $\mathbf{Y} = [\mathbf{y}_1, \ldots, \mathbf{y}_M] \in \mathbb{R}^{N \times M}$ collect the M observed signals.

As described in [140], the goal is to infer the graph topology of brain networks from the time series data $\mathbf{Y}$. Specifically, the goal is to identify a weighted, undirected, and unknown graph $\mathcal{G} = (\mathcal{V}, \mathcal{E}, \mathbf{W})$ composed of N vertices. Each vertex represents a network node. An edge between two vertices signifies the physical proximity or relationship between the corresponding nodes. The set $\mathcal{V} = \{1, \cdots, N\}$ contains N vertices, while $\mathcal{E} \subseteq \mathcal{V} \times \mathcal{V}$ represents the set of network connectivity edges. The adjacency matrix $\mathbf{W} \in \mathbb{R}^{N \times N}$ reflects the correlation between nodes, with $W_{ij} = W_{ji} \neq 0$ for any $(i, j) \in \mathcal{E}$.

The combinatorial graph Laplacian $\mathbf{L}$ of the brain network $\mathcal{G}$ is defined as [146]:

$$\mathbf{L} = \mathbf{D} - \mathbf{W}, \tag{4.1}$$

where $\mathbf{D} \triangleq \mathrm{diag}\,(\mathbf{W}\mathbf{1})$ defines the degree matrix containing the node degrees at its diagonal. $\mathbf{1}$ is an all-one vector.

The graph Laplacian, a semi-definite matrix, has positive entries on its main diagonal and non-positive entries elsewhere [52]. Through eigenvalue decomposition, $\mathbf{L}$ can be expressed as:

$$\mathbf{L} = \mathbf{U}\mathbf{\Lambda}\mathbf{U}^T, \tag{4.2}$$

where $\mathbf{\Lambda}$ is the diagonal matrix of non-negative Laplacian eigenvalues, and $\mathbf{U} = [\mathbf{u}_1, \cdots, \mathbf{u}_N]$ is the orthonormal matrix containing all the eigenvectors.

Estimating the Laplacian matrix $\mathbf{L}$ is essential to uncovering the topology of $\mathcal{G}$. Following the approach in [140] and [91], the signals $\mathbf{Y}$ are assumed band-limited over graph $\mathcal{G}$, meaning the signals are sparse in the canonical domain [140, 91, 170]. The GFT [146, 131] is used to decompose a brain time series into the orthonormal components $\mathbf{U}$ in the Laplacian $\mathbf{L}$ [52].

For any $m \in \{1, \cdots, M\}$, the GFT of the time series $\mathbf{y}_m$, denoted by $\mathbf{s}_m$, projects $\mathbf{y}_m$ onto the spectral-domain subspace spanned by $\mathbf{U}$, as given by

$$\mathbf{s}_m = \mathbf{U}^T \mathbf{y}_m. \tag{4.3}$$

Due to the band-limited nature of the observed signal $\mathbf{y}_m$, $\mathbf{s}_m$ is a sparse vector, capturing the key characteristics of $\mathbf{y}_m$ in the frequency domain. The band-limited signal is expressed as

$$\mathbf{y}_m = \mathbf{U}\mathbf{s}_m. \tag{4.4}$$

Let $\mathbf{S} = [\mathbf{s}_1, \cdots, \mathbf{s}_M] \in \mathbb{R}^{N \times M}$ collect all $\mathbf{s}_m \in \mathbb{R}^{N \times 1}$, $m \in \{1, \cdots, M\}$. From (4.4), it has

$$\mathbf{Y} = \mathbf{U}\mathbf{S}. \tag{4.5}$$

With the sparsity of $\mathbf{s}_m, m \in \{1, \cdots, M\}$, $\mathbf{S} \in \mathcal{B}_K$ is set as a K-block sparse matrix with multiple all-zero row-vectors. K accounts for the bandwidth of the frequency-domain representation of the observed band-limited graph signals $\mathbf{Y}$, which can be obtained empirically in prior, or enumerated to find its proper value. $\mathcal{B}_K$ collects all K-block sparse matrices [53]:

$$\mathcal{B}_K \triangleq \{\mathbf{S} \in \mathbb{R}^{N \times M}, \mathbf{S}(i, :) = \mathbf{0}, \forall i \notin \mathcal{K} \subseteq \mathcal{V}, K = |\mathcal{K}|\}, \tag{4.6}$$

where $\mathbf{S}(i, :)$ is the i-th row of $\mathbf{S}$, and $\mathcal{K} \subseteq \mathcal{V}$ has the cardinality of K.

4.3 Proposed Alternating Optimization for Graph Learning

The Laplacian matrix $\mathbf{L}$ is estimated and, in turn, the topological knowledge of graph $\mathcal{G}$ substantiates the observation $\mathbf{Y}$. Given the band-limitedness of $\mathbf{Y}$, the problem can be cast as

$$\min_{\mathbf{L}, \mathbf{U} \in \mathbb{R}^{N \times N}, \mathbf{S} \in \mathbb{R}^{N \times M}} \|\mathbf{Y} - \mathbf{U}\mathbf{S}\|_F^2 + \beta f(\mathbf{L}, \mathbf{Y}) \tag{4.7a}$$

$$\text{s.t.} \quad \mathbf{U}\mathbf{U}^T = \mathbf{I}_N, \tag{4.7b}$$

$$\mathbf{S} \in \mathcal{B}_K, \tag{4.7c}$$

$$\mathbf{L} = \mathbf{U}\mathbf{\Lambda}\mathbf{U}^T, \ \mathbf{L} \in \mathbb{L}, \tag{4.7d}$$

$$\mathbf{u}_1 = \frac{1}{\sqrt{N}}\mathbf{1}. \tag{4.7e}$$

The first term of the objective (4.7a) accounts for data fidelity by incorporating a quadratic loss that penalizes any disparity among $\mathbf{US}$ and $\mathbf{Y}$. The second term of (4.7a) is a regularization function [44, 140]. β is an adjustable weighting coefficient of the regularizer.

Constraint (4.7b) ensures $\mathbf{U}$ to be a unitary matrix complying with (4.2). Constraint (4.7c) imposes the K-block sparsity of $\mathbf{S}$ in (4.6). Constraint (4.7d) guarantees that $\mathbf{L}$ refers to a Laplacian matrix that satisfies the necessary conditions and properties, and $\mathbb{L}$ collects all valid candidates to $\mathbf{L}$ [52], i.e.,

$$\mathbb{L} = \{\mathbf{L} \succeq \mathbf{0} | \mathbf{L}\mathbf{1} = \mathbf{0}, L_{ij} = L_{ji} \le 0, i \ne j\}, \tag{4.8}$$

where $\mathbf{0}$ refers to an all-zero vector. According to $\mathbf{L}\mathbf{1} = \mathbf{0}$ in (4.8), it can be concluded that 0 is one eigenvalue of $\mathbf{L}$ corresponding to $\mathbf{u}_1 = \frac{1}{\sqrt{N}}\mathbf{1}$, i.e., the first column of $\mathbf{U}$; see (4.7e).

Given the *a-priori* statistical knowledge of $\mathbf{L}$, the regularizer $f(\mathbf{L}, \mathbf{Y})$ can effectively capture and reflect the intended characteristics of the graph topology. Suppose that $\mathbf{L}$ follows an exponential distribution [52]. The maximum a-posteriori (MAP) estimate of $\mathbf{L}$ is written as [52]

$$\min_{\mathbf{L}} \ \mathrm{tr}(\mathbf{L}\mathbf{Y}\mathbf{Y}^T) - \log|\mathbf{L}| + \alpha \|\mathrm{vec}(\mathbf{L})\|_1, \ \ \mathrm{s.t.} \ \ \mathbf{L} \in \mathbb{L}. \tag{4.9}$$

In this sense, $f(\mathbf{L}, \mathbf{Y}) = \mathrm{tr}(\mathbf{L}\mathbf{Y}\mathbf{Y}^T) - \log|\mathbf{L}| + \alpha \|\mathrm{vec}(\mathbf{L})\|_1$ provides a reasonable regularization in (4.7a). Here, $\|\cdot\|_1$ stands for the ℓ_1-norm and α is a tunable regularization parameter.

Based on the non-positivity and the structural constraint (i.e., $\mathbf{L}\mathbf{1} = \mathbf{0}$) of $\mathbf{L}$, it obtains $\|\mathrm{vec}(\mathbf{L})\|_1 = 2\mathrm{tr}(\mathbf{L})$. By using the linearity of trace, $f(\mathbf{L}, \mathbf{Y})$ can be rewritten as

$$f(\mathbf{L}, \mathbf{Y}) = \mathrm{tr}\left(\mathbf{L}\mathbf{Y}\mathbf{Y}^T + 2\alpha\mathbf{L}\right) - \log|\mathbf{L}| = \mathrm{tr}\left(\mathbf{L}\mathbf{T}\right) - \log|\mathbf{L}|, \tag{4.10}$$

where $\mathbf{T} = \mathbf{Y}\mathbf{Y}^T + 2\alpha\mathbf{I}$ for notational simplicity.

As $f(\mathbf{L}, \mathbf{Y}) = \mathrm{tr}\left(\mathbf{L}\mathbf{T}\right) - \log|\mathbf{L}|$ is intractable due to the singular pseudo-determinant of $|\mathbf{L}|$, the equivalence of $\log|\mathbf{L}|$ and $\log\det(\mathbf{L}+\mathbf{J})$ with $\mathbf{J} = \mathbf{u}_1\mathbf{u}_1^T = \frac{1}{N}\mathbf{1}\mathbf{1}^T$ [52, Prop. 1] is explored and $f(\mathbf{L}, \mathbf{Y})$ is rewritten as

$$f(\mathbf{L}, \mathbf{Y}) = \mathrm{tr}\left(\mathbf{L}\mathbf{T}\right) - \log\det(\mathbf{L}+\mathbf{J}). \tag{4.11}$$

By exploiting the orthonormality of $\mathbf{U}$ in (4.7b), $\|\mathbf{Y} - \mathbf{US}\|_F^2 = \|\mathbf{U}^T\mathbf{Y} - \mathbf{S}\|_F^2$ is obtained. Then, (4.7) can be rewritten as

$$\min_{\mathbf{L},\mathbf{U}\in\mathbb{R}^{N\times N},\mathbf{S}\in\mathbb{R}^{N\times M}} \|\mathbf{U}^T\mathbf{Y} - \mathbf{S}\|_F^2 + \beta f(\mathbf{L},\mathbf{Y})$$

$$\text{s.t. } (4.7\text{b}) - (4.7\text{e}). \tag{4.12}$$

Problem (4.12) is non-convex because of the non-convex nature of the objective function $f(\mathbf{L},\mathbf{Y})$ and the coupling of the variables in (4.7a), the orthonormality in (4.7b), and the sparsity in (4.7c). Since both $\mathbf{U}$ and $\mathbf{S}$ are unknown, (4.12) is reorganized as

$$\min_{\mathbf{L},\mathbf{U}\in\mathbb{R}^{N\times N}} \left(\min_{\mathbf{S}\in\mathcal{B}_K} \sum_{i=1}^{N} \|\mathbf{u}_i^T\mathbf{Y} - \mathbf{S}(i,:)\|_2^2 + \beta f(\mathbf{L},\mathbf{Y}) \right)$$

$$\text{s.t. } (4.7\text{b}) - (4.7\text{e}), \tag{4.13}$$

which, based on the definition of $\mathbf{S}$ (i.e., $\mathbf{S}(i,:) = \mathbf{0}$), can be further rewritten as

$$\min_{\mathbf{L},\mathbf{U}\in\mathbb{R}^{N\times N},\mathcal{K}} \left(\min_{\mathbf{S}\in\mathcal{B}_K} \sum_{i\in\mathcal{K}} \|\mathbf{u}_i^T\mathbf{Y} - \mathbf{S}(i,:)\|_2^2 + \sum_{i\notin\mathcal{K}} \|\mathbf{u}_i^T\mathbf{Y}\|_2^2 \right) + \beta f(\mathbf{L},\mathbf{Y}),$$

$$\tag{4.14}$$

$$\text{s.t. } (4.7\text{b}) - (4.7\text{e}).$$

By closely assessing the objective of (4.14), it can be found that the optimal index set, $\mathcal{K}$, collects the indices to the K largest entries of $\{\|\mathbf{u}_i^T\mathbf{Y}\|\}_i^N$, and satisfies

$$\mathbf{S}(i,:) = \begin{cases} \mathbf{u}_i^T\mathbf{Y}, & \text{if } i \in \mathcal{K}; \\ \mathbf{0}, & \text{if } i \notin \mathcal{K}. \end{cases} \tag{4.15}$$

As a result, only the $N - K$ smallest entries of $\{\|\mathbf{u}_i^T\mathbf{Y}\|\}_i^N$ remain in the objective of (4.14). The first term of (4.7a) is rewritten as

$$\|\mathbf{Y} - \mathbf{US}\|_F^2 = \|\mathbf{Y} - \mathbf{U}_\mathcal{K}\mathbf{S}_\mathcal{K}\|_F^2 = \|(\mathbf{I} - \mathbf{U}_\mathcal{K}\mathbf{U}_\mathcal{K}^T)\mathbf{Y}\|_F^2, \tag{4.16}$$

where $\mathbf{U}_\mathcal{K}$ is the matrix collating the column-vectors of $\mathbf{U}$ indexed by $\mathcal{K}$, and $\mathbf{S}_\mathcal{K}$ is the matrix collating the row-vectors of $\mathbf{S}$ indexed by $\mathcal{K}$. The second component in the objective of problem (4.12), i.e., $f(\mathbf{L},\mathbf{Y})$, can be reformulated by considering the following two cases in regards to constraint (4.7e):

1. *In the case of* $\mathbf{u}_1 \notin \mathbf{U}_\mathcal{K}$, the eigenvectors of $\mathbf{L}$, i.e., $\mathbf{U}$, can be arranged as $[\mathbf{u}_1, \mathbf{U}_\mathcal{K}, \mathbf{U}_{\mathcal{K}^c\setminus\{1\}}]$. By performing eigenvalue decomposition, $\mathbf{L}$ can be formulated to

$$\mathbf{L} = [\mathbf{u}_1, \mathbf{U}_\mathcal{K}, \mathbf{U}_{\mathcal{K}^c\setminus\{1\}}] \boldsymbol{\Phi}_1 [\mathbf{u}_1, \mathbf{U}_\mathcal{K}, \mathbf{U}_{\mathcal{K}^c\setminus\{1\}}]^T$$

$$= [\mathbf{U}_\mathcal{K}, \mathbf{U}_{\mathcal{K}^c\setminus\{1\}}] \boldsymbol{\Phi}_2 [\mathbf{U}_\mathcal{K}, \mathbf{U}_{\mathcal{K}^c\setminus\{1\}}]^T. \tag{4.17}$$

By plugging (4.17) and $\mathbf{J} = \frac{1}{N}\mathbf{1}\mathbf{1}^T$, $\log\det(\mathbf{L} + \mathbf{J})$ can be rewritten as

$\log\det(\mathbf{L} + \mathbf{J})$

$$= \log\det\left(\left[\frac{1}{\sqrt{N}}, \mathbf{U}_{\mathcal{K}}, \mathbf{U}_{\mathcal{K}^c\setminus\{1\}}\right]\begin{bmatrix} 1 & \\ & \mathbf{\Phi}_2 \end{bmatrix}\left[\frac{1}{\sqrt{N}}, \mathbf{U}_{\mathcal{K}}, \mathbf{U}_{\mathcal{K}^c\setminus\{1\}}\right]^T\right)$$

$$= \log\det\left(\mathrm{blkdiag}\left(1, \mathbf{\Lambda}_{\mathcal{K}}, \mathbf{\Lambda}_{\mathcal{K}^c\setminus\{1\}}\right)\right)$$

$$= \log\det(\mathbf{\Lambda}_{\mathcal{K}}) + \log\det(\mathbf{\Lambda}_{\mathcal{K}^c\setminus\{1\}}). \tag{4.18}$$

Then, $f(\mathbf{L}, \mathbf{Y})$ can be formulated to

$$f(\mathbf{L}, \mathbf{Y}) = \mathrm{tr}\left(\mathbf{U}_{\mathcal{K}}\mathbf{\Lambda}_{\mathcal{K}}\mathbf{U}_{\mathcal{K}}^T\mathbf{T}\right)$$
$$+ \mathrm{tr}\left(\mathbf{U}_{\mathcal{K}^c\setminus\{1\}}\mathbf{\Lambda}_{\mathcal{K}^c\setminus\{1\}}\mathbf{U}_{\mathcal{K}^c\setminus\{1\}}^T\mathbf{T}\right)$$
$$- \log\det(\mathbf{\Lambda}_{\mathcal{K}}) - \log\det(\mathbf{\Lambda}_{\mathcal{K}^c\setminus\{1\}}), \tag{4.19}$$

where $\mathbf{\Phi}_1 = \mathrm{blkdiag}\left(0, \mathbf{\Lambda}_{\mathcal{K}}, \mathbf{\Lambda}_{\mathcal{K}^c\setminus\{1\}}\right)$ with $\mathbf{\Lambda}_{\mathcal{K}} \succeq 0$ and $\mathbf{\Lambda}_{\mathcal{K}} \in \mathbb{R}^{K\times K}$ is a block-diagonal matrix and collects all the eigenvalues associated with the eigenvectors $[\mathbf{u}_1, \mathbf{U}_{\mathcal{K}}, \mathbf{U}_{\mathcal{K}^c\setminus\{1\}}]$ of $\mathbf{L}$; and $\mathbf{\Phi}_2 = \mathrm{blkdiag}\left(\mathbf{\Lambda}_{\mathcal{K}}, \mathbf{\Lambda}_{\mathcal{K}^c\setminus\{1\}}\right)$ with $\mathbf{\Lambda}_{\mathcal{K}^c\setminus\{1\}} \succeq 0$ and $\mathbf{\Lambda}_{\mathcal{K}^c\setminus\{1\}} \in \mathbb{R}^{(N-K-1)\times(N-K-1)}$.

2. *In the case of* $\mathbf{u}_1 \in \mathbf{U}_{\mathcal{K}}$, $\mathbf{U}$ *can be arranged as* $[\mathbf{u}_1, \mathbf{U}_{\mathcal{K}\setminus\{1\}}, \mathbf{U}_{\mathcal{K}^c}]$. *Then, (4.16) is rewritten as*

$$\left\|(\mathbf{I} - \mathbf{U}_{\mathcal{K}}\mathbf{U}_{\mathcal{K}}^T)\mathbf{Y}\right\|_F^2 = \left\|\left(\mathbf{I} - \mathbf{U}_{\mathcal{K}\setminus\{1\}}\mathbf{U}_{\mathcal{K}\setminus\{1\}}^T - \mathbf{u}_1\mathbf{u}_1^T\right)\mathbf{Y}\right\|_F^2$$

$$\overset{(a)}{=} \mathrm{tr}\left[\left(\mathbf{I} - \mathbf{U}_{\mathcal{K}\setminus\{1\}}\mathbf{U}_{\mathcal{K}\setminus\{1\}}^T\right)\mathbf{Y}\mathbf{Y}^T\left(\mathbf{I} - \mathbf{U}_{\mathcal{K}\setminus\{1\}}\mathbf{U}_{\mathcal{K}\setminus\{1\}}^T\right)^T\right.$$
$$\left. - \mathbf{u}_1\mathbf{u}_1^T\mathbf{Y}\mathbf{Y}^T\mathbf{u}_1\mathbf{u}_1^T + \mathbf{u}_1\mathbf{u}_1^T\mathbf{U}_{\mathcal{K}\setminus\{1\}}\mathbf{U}_{\mathcal{K}\setminus\{1\}}^T\mathbf{Y}\mathbf{Y}^T\right]$$

$$\overset{(b)}{=} \left\|\left(\mathbf{I} - \mathbf{U}_{\mathcal{K}\setminus\{1\}}\mathbf{U}_{\mathcal{K}\setminus\{1\}}^T\right)\mathbf{Y}\right\|_F^2 - \left\|\mathbf{u}_1\mathbf{u}_1^T\mathbf{Y}\right\|_F^2, \tag{4.20}$$

where $\mathbf{u}_1\mathbf{u}_1^T\mathbf{U}_{\mathcal{K}\setminus\{1\}}\mathbf{U}_{\mathcal{K}\setminus\{1\}}^T\mathbf{Y}\mathbf{Y}^T = 0$ in (a) since $\mathbf{u}_1^T\mathbf{U}_{\mathcal{K}\setminus\{1\}} = 0$, and $\mathbf{u}_1\mathbf{u}_1^T\mathbf{Y}$ is constant in (b).

Then, $\mathbf{L}$ can be rewritten as

$$\mathbf{L} = [\mathbf{u}_1, \mathbf{U}_{\mathcal{K}\setminus\{1\}}, \mathbf{U}_{\mathcal{K}^c}]\mathbf{\Phi}_1[\mathbf{u}_1, \mathbf{U}_{\mathcal{K}\setminus\{1\}}, \mathbf{U}_{\mathcal{K}^c}]^T$$

$$= [\mathbf{U}_{\mathcal{K}\setminus\{1\}}, \mathbf{U}_{\mathcal{K}^c}]\mathbf{\Phi}_2[\mathbf{U}_{\mathcal{K}\setminus\{1\}}, \mathbf{U}_{\mathcal{K}^c}]^T. \tag{4.21}$$

Likewise, $f(\mathbf{L}, \mathbf{Y})$ is rewritten as

$$f(\mathbf{L}, \mathbf{Y}) = \mathrm{tr}\left(\mathbf{U}_{\mathcal{K}\setminus\{1\}}\mathbf{\Lambda}_{\mathcal{K}\setminus\{1\}}\mathbf{U}_{\mathcal{K}\setminus\{1\}}^T\mathbf{T}\right) + \mathrm{tr}\left(\mathbf{U}_{\mathcal{K}}\mathbf{\Lambda}_{\mathcal{K}}\mathbf{U}_{\mathcal{K}}^T\mathbf{T}\right)$$
$$- \log\det(\mathbf{\Lambda}_{\mathcal{K}\setminus\{1\}}) - \log\det(\mathbf{\Lambda}_{\mathcal{K}}), \tag{4.22}$$

where, with a slight abuse of notation, $\boldsymbol{\Phi}_1 = \text{blkdiag}\left(0, \boldsymbol{\Lambda}_{\mathcal{K}\backslash\{1\}}, \boldsymbol{\Lambda}_{\mathcal{K}^c}\right)$ with $\boldsymbol{\Lambda}_{\mathcal{K}\backslash\{1\}} \succeq \mathbf{0}$ and $\boldsymbol{\Lambda}_{\mathcal{K}\backslash\{1\}} \in \mathbb{R}^{(K-1)\times(K-1)}$; and $\boldsymbol{\Phi}_2 = \text{blkdiag}\left(\boldsymbol{\Lambda}_{\mathcal{K}\backslash\{1\}}, \boldsymbol{\Lambda}_{\mathcal{K}^c}\right)$ with $\boldsymbol{\Lambda}_{\mathcal{K}^c} \succeq \mathbf{0}$ with $\boldsymbol{\Lambda}_{\mathcal{K}^c} \in \mathbb{R}^{(N-K)\times(N-K)}$.

To unify the presentation and the follow-on discussions of the two cases, the following definition is introduced

1. If $\mathbf{u}_1 \notin \mathbf{U}_{\mathcal{K}}$, then $\mathbf{V}_1 = \mathbf{U}_{\mathcal{K}}$, $\mathbf{V}_2 = \mathbf{U}_{\mathcal{K}^c\backslash\{1\}}$, $\boldsymbol{\Lambda}_1 = \boldsymbol{\Lambda}_{\mathcal{K}}, \boldsymbol{\Lambda}_2 = \boldsymbol{\Lambda}_{\mathcal{K}^c\backslash\{1\}}, \mathbf{I}_1 \in \mathbb{R}^{K\times K}$, and $\mathbf{I}_2 \in \mathbb{R}^{(N-K-1)\times(N-K-1)}$;

2. If $\mathbf{u}_1 \in \mathbf{U}_{\mathcal{K}}$, then $\mathbf{V}_1 = \mathbf{U}_{\mathcal{K}\backslash\{1\}}$, $\mathbf{V}_2 = \mathbf{U}_{\mathcal{K}^c}$, $\boldsymbol{\Lambda}_1 = \boldsymbol{\Lambda}_{\mathcal{K}\backslash\{1\}}, \boldsymbol{\Lambda}_2 = \boldsymbol{\Lambda}_{\mathcal{K}^c}, \mathbf{I}_1 \in \mathbb{R}^{(K-1)\times(K-1)}$, and $\mathbf{I}_2 \in \mathbb{R}^{(N-K)\times(N-K)}$.

As a result, $\mathbf{L}$ can be represented as

$$\mathbf{L} = [\mathbf{V}_1, \mathbf{V}_2]\,\boldsymbol{\Phi}_2\,[\mathbf{V}_1, \mathbf{V}_2]^T. \tag{4.23}$$

Since $[\mathbf{V}_1, \mathbf{V}_2]^T\mathbf{u}_1 = 0$, $\mathbf{L}\mathbf{1} = 0$ in (4.8) is preserved in (4.23).

Finally, with the unified presentation (4.23) capturing both cases of $\mathbf{u}_1 \in \mathcal{K}$ and $\mathbf{u}_1 \notin \mathcal{K}$, problem (4.7) is transformed equivalently to

$$\min_{\mathbf{V}_1,\mathbf{V}_2,\boldsymbol{\Lambda}_1,\boldsymbol{\Lambda}_2} \left\|\left(\mathbf{I} - \mathbf{V}_1\mathbf{V}_1^T\right)\mathbf{Y}\right\|_F^2 + \beta\left[\text{tr}\left(\mathbf{V}_1\boldsymbol{\Lambda}_1\mathbf{V}_1^T\mathbf{T}\right) + \right.$$
$$\left. \text{tr}\left(\mathbf{V}_2\boldsymbol{\Lambda}_2\mathbf{V}_2^T\mathbf{T}\right) - \log\det\left(\boldsymbol{\Lambda}_1\right) - \log\det\left(\boldsymbol{\Lambda}_2\right)\right] \tag{4.24a}$$

$$\text{s.t.}\quad \mathbf{V}_1^T\mathbf{V}_1 = \mathbf{I}_1, \mathbf{V}_2^T\mathbf{V}_2 = \mathbf{I}_2, \mathbf{V}_1^T\mathbf{V}_2 = \mathbf{0}, \tag{4.24b}$$

$$\mathbf{u}_1^T\mathbf{V}_1 = \mathbf{0}, \mathbf{u}_1^T\mathbf{V}_2 = \mathbf{0}, \tag{4.24c}$$

$$\boldsymbol{\Lambda}_1 \succeq \mathbf{0}, \boldsymbol{\Lambda}_2 \succeq \mathbf{0}, \tag{4.24d}$$

$$\mathbf{I} \odot [\mathbf{V}_1\boldsymbol{\Lambda}_1\mathbf{V}_1^T + \mathbf{V}_2\boldsymbol{\Lambda}_2\mathbf{V}_2^T] \geq 0, \tag{4.24e}$$

$$\mathbf{A} \odot [\mathbf{V}_1\boldsymbol{\Lambda}_1\mathbf{V}_1^T + \mathbf{V}_2\boldsymbol{\Lambda}_2\mathbf{V}_2^T] \leq 0, \tag{4.24f}$$

where (4.24b) is from (4.7b), and (4.24c)–(4.24f) correspond to (4.8).

Next, AO is carried out to iteratively solve problem (4.24) by optimizing $[\mathbf{V}_1, \mathbf{V}_2]$ and $[\boldsymbol{\Lambda}_1, \boldsymbol{\Lambda}_2]$ in an alternating manner. Finally, given the convergent $\boldsymbol{\Lambda}_1$ and $\boldsymbol{\Lambda}_2$, the Laplacian matrix is inferred by $\mathbf{L}$ using (4.23).

4.3.1 Estimation of Graph Fourier Basis

Given $\boldsymbol{\Lambda}_1$ and $\boldsymbol{\Lambda}_2$, problem (4.24) is reduced to

$$\min_{\mathbf{V}_1,\mathbf{V}_2} \left\|\left(\mathbf{I} - \mathbf{V}_1\mathbf{V}_1^T\right)\mathbf{Y}\right\|_F^2 + \beta\left[\text{tr}\left(\mathbf{V}_1\boldsymbol{\Lambda}_1\mathbf{V}_1^T\mathbf{T}\right) + \text{tr}\left(\mathbf{V}_2\boldsymbol{\Lambda}_2\mathbf{V}_2^T\mathbf{T}\right)\right] \tag{4.25a}$$

$$\text{s.t.}\quad \mathbf{V}_1^T\mathbf{V}_1 = \mathbf{I}_1, \mathbf{V}_2^T\mathbf{V}_2 = \mathbf{I}_2, \mathbf{V}_1^T\mathbf{V}_2 = \mathbf{0}, \tag{4.25b}$$

$$\mathbf{u}_1^T\mathbf{V}_1 = \mathbf{0}, \mathbf{u}_1^T\mathbf{V}_2 = \mathbf{0}, \tag{4.25c}$$

which has orthonormal vector variables (or, in other words, the feasible solution region is on the Stiefel manifold) and, therefore, can be solved using Stiefel manifold dual gradient descent [129], as described below.

The Lagrangian function of (4.25) is defined as

$$\mathcal{L}\left(\mathbf{V}_1, \mathbf{V}_2, \mathbf{\Psi}_1, \mathbf{\Psi}_2, \mathbf{\Psi}_3, \mathbf{\Psi}_4, \mathbf{\Psi}_5\right)$$
$$= \left\|(\mathbf{I} - \mathbf{V}_1\mathbf{V}_1^T)\,\mathbf{Y}\right\|_F^2 + \beta\operatorname{tr}\left(\mathbf{V}_1\mathbf{\Lambda}_1\mathbf{V}_1^T\mathbf{T}\right)$$
$$+ \beta\operatorname{tr}\left(\mathbf{V}_2\mathbf{\Lambda}_2\mathbf{V}_2^T\mathbf{T}\right) - \frac{1}{2}\operatorname{tr}(\mathbf{\Psi}_1^T(\mathbf{V}_1^T\mathbf{V}_1 - \mathbf{I}_1))$$
$$- \frac{1}{2}\operatorname{tr}(\mathbf{\Psi}_2^T(\mathbf{V}_2^T\mathbf{V}_2 - \mathbf{I}_2)) - \frac{1}{2}\operatorname{tr}(\mathbf{\Psi}_3^T\mathbf{V}_1^T\mathbf{V}_2)$$
$$- \frac{1}{2}\operatorname{tr}(\mathbf{\Psi}_4^T\mathbf{u}_1^T\mathbf{V}_1) - \frac{1}{2}\operatorname{tr}(\mathbf{\Psi}_5^T\mathbf{u}_1^T\mathbf{V}_2), \tag{4.26}$$

where $\mathbf{\Psi}_1, \cdots, \mathbf{\Psi}_5$ are the Lagrange multipliers corresponding to the five conditions in (4.24).

Considering the Karush–Kuhn–Tucker (KKT) conditions, (4.26) regarding $\mathbf{V}_1$ is differentiated and then set to zero, as given by

$$\nabla_{\mathbf{V}_1}\mathcal{L} = \nabla\mathcal{F}(\mathbf{V}_1) - \mathbf{V}_1\mathbf{\Psi}_1 - \frac{1}{2}\mathbf{V}_2\mathbf{\Psi}_3 - \frac{1}{2}\mathbf{u}_1\mathbf{\Psi}_4 = 0, \tag{4.27}$$

where $\mathcal{F}(\mathbf{V}_1, \mathbf{V}_2) = \left\|(\mathbf{I} - \mathbf{V}_1\mathbf{V}_1^T)\,\mathbf{Y}\right\|_F^2 + \beta\operatorname{tr}\left(\mathbf{V}_1\mathbf{\Lambda}_1\mathbf{V}_1^T\mathbf{T}\right) + \beta\operatorname{tr}\left(\mathbf{V}_2\mathbf{\Lambda}_2\mathbf{V}_2^T\mathbf{T}\right)$.

By left multiplying $\mathbf{V}_1^T$ to both sides of (4.27), it has

$$\mathbf{V}_1^T\nabla\mathcal{F}(\mathbf{V}_1) - \mathbf{V}_1^T\mathbf{V}_1\mathbf{\Psi}_1 - \frac{1}{2}\mathbf{V}_1^T\mathbf{V}_2\mathbf{\Psi}_3 - \frac{1}{2}\mathbf{V}_1^T\mathbf{u}_1\mathbf{\Psi}_4 = 0. \tag{4.28}$$

Considering (4.24b), it obtains

$$\mathbf{\Psi}_1 = \mathbf{V}_1^T\nabla\mathcal{F}(\mathbf{V}_1). \tag{4.29}$$

By left multiplying $\mathbf{V}_2^T$ to both sides of (4.27) and applying (4.24b), it has

$$\mathbf{\Psi}_3 = 2\nabla\mathcal{F}(\mathbf{V}_1)^T\mathbf{V}_2. \tag{4.30}$$

By left multiplying $\mathbf{u}_1^T$ to both sides of (4.27) and applying (4.24b), it has

$$\mathbf{\Psi}_4 = 2\mathbf{u}_1^T\nabla\mathcal{F}(\mathbf{V}_1). \tag{4.31}$$

Since the constraint $\mathbf{V}_1^T\mathbf{V}_1 = \mathbf{I}_1$ in (4.25) is symmetric, $\mathbf{\Psi}_1$ is symmetric and therefore $\mathbf{\Psi}_1 = \nabla\mathcal{F}(\mathbf{V}_1)^T\mathbf{V}_1$. The gradient in (4.27) can be reformulated to

$$\nabla_{\mathbf{V}_1}\mathcal{L} = \nabla\mathcal{F}(\mathbf{V}_1) - \mathbf{V}_1\nabla\mathcal{F}(\mathbf{V}_1)^T\mathbf{V}_1 - \frac{1}{2}\mathbf{V}_2\mathbf{V}_2^T\nabla\mathcal{F}(\mathbf{V}_1) - \frac{1}{2}\mathbf{u}_1\mathbf{u}_1^T\nabla\mathcal{F}(\mathbf{V}_1).$$
$$\tag{4.32}$$

Likewise,

$$\boldsymbol{\Psi}_2 = 2\mathbf{V}_2^T \nabla \mathcal{F}(\mathbf{V}_2); \ \boldsymbol{\Psi}_5 = 2\mathbf{u}_1^T \nabla \mathcal{F}(\mathbf{V}_2). \tag{4.33}$$

The gradient of the Lagrangian regarding $\mathbf{V}_2$ can be obtained by swapping $\mathbf{V}_1$ and $\mathbf{V}_2$ in (4.32).

With the Stiefel manifold dual gradient descent, problem (4.25) can be solved by iteratively updating the Lagrange multipliers with (4.29)–(4.31) and (4.33) and the variables $\mathbf{V}_1$ and $\mathbf{V}_2$ with the right-hand scaled gradient projection method [129]:

$$\mathbf{V}_1 \leftarrow \pi(\mathbf{V}_1 - \tau_1 \nabla_{\mathbf{V}_1} \mathcal{L}(\mathbf{V}_1) \mathcal{A}_1(\mu, \tau_1)), \tag{4.34}$$

where $\pi(\cdot)$ is the projection operator, i.e., $\pi(\mathbf{V}_1) = \mathbf{U}\mathbf{I}_1\mathbf{V}^T$ if $\mathbf{V}_1 = \mathbf{U}\boldsymbol{\Sigma}\mathbf{V}^T$ by singular value decomposition (SVD) [129]; $\mathcal{A}_1(\mu, \tau_1)$ is a scaling matrix with $\mu \in (0,1)$, i.e.,

$$\mathcal{A}_1(\mu, \tau_1) = (\mathbf{I}_1 + \mu\tau_1 \mathbf{V}_1^T \nabla_{\mathbf{V}_1} \mathcal{L}(\mathbf{V}_1))^{-1}, \tag{4.35}$$

and τ_1 is the step size and given by

$$\tau_1 = \begin{cases} \dfrac{\|\mathbf{V}_1 - \mathbf{V}_1'\|_F^2}{\langle \mathbf{V}_1 - \mathbf{V}_1', \nabla_{\mathbf{V}_1}\mathcal{L}(\mathbf{V}_1) - \nabla_{\mathbf{V}_1}\mathcal{L}(\mathbf{V}_1')\rangle}, & \text{in odd-numbered iterations,} \\[2ex] \dfrac{\langle \mathbf{V}_1 - \mathbf{V}_1', \nabla_{\mathbf{V}_1}\mathcal{L}(\mathbf{V}_1) - \nabla_{\mathbf{V}_1}\mathcal{L}(\mathbf{V}_1')\rangle}{\|\nabla_{\mathbf{V}_1}\mathcal{L}(\mathbf{V}_1) - \nabla_{\mathbf{V}_1}\mathcal{L}(\mathbf{V}_1')\|_F^2}, & \text{in even-numbered iterations.} \end{cases} \tag{4.36}$$

Here, $\mathbf{V}_1'$ is the counterpart of $\mathbf{V}_1$ at the previous iteration. $\tau_1 \in [\tau_{\min}, \tau_{\max}]$ with $\tau_{\min} = 10^{-5}$ and $\tau_{\max} = 10^5$ being the minimum and maximum values of step-size, respectively.

The right-hand scaled gradient projection of $\mathbf{V}_2$ can be obtained by replacing $\mathbf{V}_1$ and $\mathcal{A}_1(\mu, \tau_1)$ with $\mathbf{V}_2$ and $\mathcal{A}_2(\mu, \tau_2)$, respectively, in (4.34). Here, the scaling matrix $\mathcal{A}_2(\mu, \tau_2)$ can be obtained by replacing $\mathbf{I}_1$ and $\mathbf{V}_1$ with $\mathbf{I}_2$ and $\mathbf{V}_2$, respectively, in (4.35). The step-size $\tau_2 \in [\tau_{\min}, \tau_{\max}]$ can be obtained by replacing $\mathbf{V}_1$ and $\mathbf{V}_1'$ with $\mathbf{V}_2$ and $\mathbf{V}_2'$, respectively, in (4.36). $\mathbf{V}_2'$ is the counterpart of $\mathbf{V}_2$ in the previous iteration.

4.3.2 Estimation of Eigenvalues

Given $[\mathbf{V}_1, \mathbf{V}_2]$, problem (4.24) is rewritten as

$$\min_{\boldsymbol{\Lambda}_1, \boldsymbol{\Lambda}_2} \operatorname{tr}\left(\mathbf{V}_1 \boldsymbol{\Lambda}_1 \mathbf{V}_1^T \mathbf{T}\right) + \operatorname{tr}\left(\mathbf{V}_2 \boldsymbol{\Lambda}_2 \mathbf{V}_2^T \mathbf{T}\right) - \log\det\left(\boldsymbol{\Lambda}_1\right) - \log\det\left(\boldsymbol{\Lambda}_2\right)$$

$$\text{s.t.} \quad \boldsymbol{\Lambda}_1 \succeq 0, \boldsymbol{\Lambda}_2 \succeq 0,$$

$$\mathbf{I} \odot [\mathbf{V}_1 \boldsymbol{\Lambda}_1 \mathbf{V}_1^T + \mathbf{V}_2 \boldsymbol{\Lambda}_2 \mathbf{V}_2^T] \geq 0,$$

$$\mathbf{A} \odot [\mathbf{V}_1 \boldsymbol{\Lambda}_1 \mathbf{V}_1^T + \mathbf{V}_2 \boldsymbol{\Lambda}_2 \mathbf{V}_2^T] \leq 0. \tag{4.37}$$

Since both $\boldsymbol{\Lambda}_1$ and $\boldsymbol{\Lambda}_2$ may have zero diagonal elements, problem (4.37) is not continually differentiable and cannot be solved using CVX toolbox. By referring to [182], $\mathbf{C} = \mathbf{L}$ is introduced and the problem (4.37) is rewritten as

$$
\min_{\boldsymbol{\Lambda}_1,\boldsymbol{\Lambda}_2} \left[\operatorname{tr}\left(\boldsymbol{\Lambda}_1 \tilde{\mathbf{T}}_1\right) + \operatorname{tr}\left(\boldsymbol{\Lambda}_2 \tilde{\mathbf{T}}_2\right) - \log\det\left(\boldsymbol{\Lambda}_1\right) - \log\det\left(\boldsymbol{\Lambda}_2\right) \right]
$$

$$
\begin{aligned}
\text{s.t.} \quad & \boldsymbol{\Lambda}_1 \succeq \mathbf{0}, \boldsymbol{\Lambda}_2 \succeq \mathbf{0}, \\
& \mathbf{V}_1 \boldsymbol{\Lambda}_1 \mathbf{V}_1^T + \mathbf{V}_2 \boldsymbol{\Lambda}_2 \mathbf{V}_2^T - \mathbf{C} = \mathbf{0}, \\
& \mathbf{I} \odot \mathbf{C} \geq \mathbf{0}, \\
& \mathbf{A} \odot \mathbf{C} \leq \mathbf{0}, \quad\quad\quad\quad\quad\quad\quad\quad (4.38)
\end{aligned}
$$

which can be addressed by utilizing ADMM; see Appendix 4.5.

Algorithm 1 summarizes the proposed AO-based algorithm that solves problem (4.24). The algorithm runs until the objective of (4.24) converges with the accuracy of ξ. Each of the AO iterations starts by running the Stiefel manifold dual gradient descent till convergence, followed by the ADMM with $\boldsymbol{\Lambda}_1'$, $\boldsymbol{\Lambda}_2'$, $\mathbf{C}'$, and $\mathbf{Z}'$ being the results of $\boldsymbol{\Lambda}_1$, $\boldsymbol{\Lambda}_2$, $\mathbf{C}$, and $\mathbf{Z}$ in the previous iteration. Their respective convergence criteria are

$$
\|\nabla_{\mathbf{V}_1}\mathcal{L}(\mathbf{V}_1)\|_F \leq \xi_1 \text{ and } \|\nabla_{\mathbf{V}_2}\mathcal{L}(\mathbf{V}_2)\|_F \leq \xi_1; \quad\quad (4.39)
$$

$$
\|\mathbf{C} - \mathbf{C}'\| / \|\mathbf{C}'\| \leq \xi_2 \text{ and } \|\mathbf{Z} - \mathbf{Z}'\| / \|\mathbf{Z}'\| \leq \xi_2, \quad\quad (4.40)
$$

where ξ_1 and ξ_2 are the predefined thresholds. For example, in experiments, it takes $\xi = \xi_1 = \xi_2 = 10^{-4}$; see Section 4.4.

4.3.3 Complexity Analysis

In each iteration of the Stiefel manifold dual gradient descent, the computational cost of evaluating $\mathbf{V}_1$ is governed by the evaluation of $\nabla\mathcal{L}(\mathbf{V}_1)$ and $\mathcal{A}_1(\mu,\tau)$, which is $\mathcal{O}(N^2 K + K^3)$. Likewise, the computational cost of evaluating $\mathbf{V}_2$ is governed by the evaluation of $\nabla\mathcal{L}(\mathbf{V}_2)$ and $\mathcal{A}_2(\mu,\tau)$, which is $\mathcal{O}(NK^2 + (N - K - 1)^3)$. The total computational complexity of the Stiefel manifold dual gradient descent is $\mathcal{O}(\log(1/\xi_1)(N^2 K + NK^2 + K^3 + (N - K - 1)^3))$. In most cases, $K < N$ and the total complexity is $\mathcal{O}\left(\log(1/\xi_1)N^3\right)$.

In each iteration of the ADMM, the computational cost of evaluating primal variable $\boldsymbol{\Lambda}_1$ is governed by the matrix multiplication and eigenvalue decomposition with $\mathcal{O}(K^3)$. Likewise, the complexity of evaluating the primal variable $\boldsymbol{\Lambda}_2$ is $\mathcal{O}\left((N - K - 1)^3\right)$. The cost of evaluating the primal variable $\mathbf{C}$ is dominated by matrix multiplication with a complexity of $\mathcal{O}\left(N^2 K + N^2(N - K - 1)\right)$. The derivation of the dual variable $\mathbf{Z}$ incurs the complexity of $\mathcal{O}\left(N^2 K + N^2(N - K - 1)\right)$. Since $K < N$, the complexity of the ADMM is $\mathcal{O}(N^3)$ per iteration. The total computational cost of the ADMM is $\mathcal{O}\left(\log(1/\xi_2)N^3\right)$. As a result, the collective computational cost

---

Algorithm 1 AO-based joint optimization method for solving problem (4.24)

1: **Initialization:** ξ_1, ξ_2, μ, β, ρ, ξ.
2: **While** (4.24a) is yet to converge with the accuracy of ξ **do**

3:      **Initialization:** Randomly initialize $\mathbf{V}_1', \mathbf{V}_2' \in \mathcal{M}$
4:              as orthogonal matrices;
5:     **while** $\|\nabla_{\mathbf{V}_1}\mathcal{L}(\mathbf{V}_1)\|_F > \xi_1$ and $\|\nabla_{\mathbf{V}_2}\mathcal{L}(\mathbf{V}_2)\|_F > \xi_1$
6:     **do**
7:      Update τ_1 and τ_2 and by (4.36) ;
8:      Compute $\tau_1 = \max(\min(\tau_1, \tau_{\max}), \tau_{\min})$;
9:      Compute $\tau_2 = \max(\min(\tau_2, \tau_{\max}), \tau_{\min})$;
10:     Update $\mathbf{V}_1$ and $\mathbf{V}_2$ by (4.34);
11:     $\mathbf{V}_1' \leftarrow \mathbf{V}_1$; $\mathbf{V}_2' \leftarrow \mathbf{V}_2$;
12:     **end while.**

13:      **Initialization:** Initialize $\mathbf{C}'$ and $\mathbf{Z}'$ as identity matrices.
14:     **while** $\|\mathbf{C} - \mathbf{C}'\| / \|\mathbf{C}'\| > \xi_2$,
15:      and $\|\mathbf{Z} - \mathbf{Z}'\| / \|\mathbf{Z}'\| > \xi_2$ **do**
16:      Update $\mathbf{\Lambda}_1$ and $\mathbf{\Lambda}_2$, see (4.47);
17:      Update $\mathbf{C}$ by (4.49);
18:      Update $\mathbf{Z}$ by (4.50);
19:      $\mathbf{\Lambda}_1' \leftarrow \mathbf{\Lambda}_1$; $\mathbf{\Lambda}_2' \leftarrow \mathbf{\Lambda}_2$; $\mathbf{C}' \leftarrow \mathbf{C}$; $\mathbf{Z}' \leftarrow \mathbf{Z}$;
20:     **end while.**
21: **end while.**

---

of the joint optimization is $\mathcal{O}\bigg(N^3 \log(1/\xi) \big[\log(1/\xi_1) + \log(1/\xi_2) \big] \bigg)$, where ξ denotes the desired level of convergence.

4.4 Numerical Results on Synthetic Data

This section conducts the experiments to gauge Algorithm 1. The experiments are carried out on a laptop with an i7-8650U CPU and 16G RAM. Algorithm 1 is initialized by setting $\mathbf{C}^{(0)}$ and $\mathbf{Z}^{(0)}$ as two symmetric unit matrices and the Lagrange multiplier ρ to 1. The algorithm is stopped after 10^4 iterations or when the difference of the objective (4.24) is smaller than 10^{-5} between two consecutive iterations.

Apart from Algorithm 1, the following advanced solutions in the literature are assessed: Dong's algorithm [44], Kalofolias' algorithm [86], Sardellitti's TV algorithm [140], Sardellitti's ESA algorithm [140], and Humbert's algorithm [79].

With reference to [140], the performance metrics, including F-measure, Recall, Precision scores, and the percentage of Recovery errors, are considered. The ground-truth graph set is denoted by $\mathcal{E}_g$, while the set of recovered graphs is represented as $\mathcal{E}_r$. Precision measures the proportion of edges correctly recovered in the graphs compared to the total number of edges in the recovered graphs, i.e., Precision $= \mathcal{E}_g \cap \mathcal{E}_r / \mathcal{E}_r$. Recall measures the proportion of edges correctly recovered in the ground-truth graphs compared to the total number of edges in the ground-truth graphs, i.e., Recall $= \mathcal{E}_g \cap \mathcal{E}_r / \mathcal{E}_g$.

F-measure is calculated as the harmonic mean of Recall and Precision, providing a measure of the overall accuracy in recovering the edges:

$$\text{F-measure} = 2\frac{\text{Precision} \cdot \text{Recall}}{\text{Precision} + \text{Recall}}. \tag{4.41}$$

The correlation coefficient $\rho_{\mathbf{W}}(\mathbf{W}_0, \mathbf{W})$ (or $\rho_{\mathbf{W}}$ for brevity) between a connected ground-truth graph and its corresponding recovered graph is defined as [140]

$$\rho_{\mathbf{W}}(\mathbf{W}_0, \mathbf{W}) = \frac{\sum_{ij} W_{0ij} W_{ij}}{\sqrt{\sum_{ij} W_{0ij}^2}\sqrt{\sum_{ij} W_{ij}^2}}, \tag{4.42}$$

where $\mathbf{W}_0$ and $\mathbf{W}$ are generated from (4.1) and give the weighted adjacent matrices of the ground-truth and recovered graphs, respectively. The coefficient correlation measures the correlation between the ground-truth graph and its corresponding recovered graph. The recovery accuracy can be enhanced with a higher coefficient value of $\rho_{\mathbf{W}}$.

The estimation error, or "Error" for short, is defined as [140]

$$\text{Error} = \frac{\|\mathbf{A} - \mathbf{A}_0\|_F}{N(N-1)}, \tag{4.43}$$

where $\mathbf{A}$ and $\mathbf{A}_0$ are the binary adjacency matrices of the recovered graphs and the ground-truth graphs, respectively.

4.5   Appendix for Section 4.3

The augmented Lagrangian of (4.37) is given by

$$\mathcal{L}\left(\mathbf{\Lambda}_1, \mathbf{\Lambda}_2, \mathbf{C}, \mathbf{Z}\right) = \mathrm{tr}\left(\mathbf{\Lambda}_1 \widetilde{\mathbf{T}}_1\right) + \mathrm{tr}\left(\mathbf{\Lambda}_2 \widetilde{\mathbf{T}}_2\right) - \log\det\left(\mathbf{\Lambda}_1\right)$$
$$- \log\det\left(\mathbf{\Lambda}_2\right) + \mathrm{tr}\left(\mathbf{Z}^T \left(\mathbf{V}_1 \mathbf{\Lambda}_1 \mathbf{V}_1^T + \mathbf{V}_2 \mathbf{\Lambda}_2 \mathbf{V}_2^T - \mathbf{C}\right)\right)$$
$$+ \frac{\rho}{2}\left\|\mathbf{V}_1 \mathbf{\Lambda}_1 \mathbf{V}_1^T + \mathbf{V}_2 \mathbf{\Lambda}_2 \mathbf{V}_2^T - \mathbf{C}\right\|_F^2, \tag{4.44}$$

where ρ is the step size.

By following the standard ADMM framework, the primal variables, $\mathbf{\Lambda}_1$, $\mathbf{\Lambda}_2$, and $\mathbf{C}$, and the dual variable $\mathbf{Z}$ are updated in an alternating manner in the following steps.

4.5.1   Update $\mathbf{\Lambda}_1$ and $\mathbf{\Lambda}_2$

Based on the primal variables $\mathbf{\Lambda}_1'$, $\mathbf{\Lambda}_2'$, and $\mathbf{C}^l$, and the dual variable $\mathbf{Z}'$ obtained in the last iteration, the primal variable in the current iteration, i.e., $\mathbf{\Lambda}_1^{l+1}$, can be obtained by minimizing the augmented Lagrangian:

$$\mathbf{\Lambda}_1 = \underset{\mathbf{\Lambda}_1 \succeq \mathbf{0}}{\mathrm{argmin}} \mathcal{L}\left(\mathbf{\Lambda}_1', \mathbf{\Lambda}_2', \mathbf{C}', \mathbf{Z}'\right) \tag{4.45}$$

$$= \underset{\mathbf{\Lambda}_1 \succeq \mathbf{0}}{\mathrm{argmin}}\left\{\frac{\rho}{2}\left\|\mathbf{\Lambda}_1' + \frac{1}{\rho}(\widetilde{\mathbf{T}}_1 + \widetilde{\mathbf{Z}}_1 - \rho\widetilde{\mathbf{X}}_1)\right\|_F^2 - \log\det\left(\mathbf{\Lambda}_1'\right)\right\},$$

where $\widetilde{\mathbf{Z}}_1 = \mathbf{V}_1^T (\mathbf{Z}')^T \mathbf{V}_1$ and $\widetilde{\mathbf{X}}_1 = \mathbf{V}_1^T (\mathbf{C}' - \mathbf{V}_2 \mathbf{\Lambda}_2' \mathbf{V}_2^T)\mathbf{V}_1$.

To solve (4.45), the first-order derivative of $\frac{\rho}{2}\left\|\mathbf{\Lambda}_1' + \frac{1}{\rho}(\widetilde{\mathbf{T}}_1 + \widetilde{\mathbf{Z}}_1 - \rho\widetilde{\mathbf{X}}_1)\right\|_F^2 -$ $\log\det\left(\mathbf{\Lambda}_1'\right)$ is set to $\mathbf{0}$, i.e.,

$$\rho\mathbf{\Lambda}_1' - (\mathbf{\Lambda}_1')^{\dagger} = -\left(\widetilde{\mathbf{T}}_1 + \widetilde{\mathbf{Z}}_1 - \rho\widetilde{\mathbf{X}}_1\right). \tag{4.46}$$

Taking the orthogonal eigenvalue decomposition of the right-hand side (RHS) of (4.46) yields $\rho\mathbf{\Lambda}_1' - (\mathbf{\Lambda}_1')^{\dagger} = -\mathbf{Q}_1 \mathbf{\Xi}_1 \mathbf{Q}_1^T$, Then, by left multiplying $\mathbf{Q}_1^T$ and right multiplying $\mathbf{Q}_1$ on both sides and it obtains $\rho\widetilde{\mathbf{\Lambda}}_1' - (\widetilde{\mathbf{\Lambda}}_1')^{\dagger} = -\mathbf{\Xi}_1$ $\widetilde{\Lambda}_{1i}' = \frac{-\rho\xi_{1i} + \sqrt{\rho^2 \xi_{1i}^2 + 4\rho}}{2\rho}$. The optimal solution to (4.45) is given by

$$\mathbf{\Lambda}_1 = \mathbf{Q}_1 \widetilde{\mathbf{\Lambda}}_1' \mathbf{Q}_1^T, \tag{4.47}$$

with the diagonal matrix $\widetilde{\mathbf{\Lambda}}_1' = \mathrm{diag}(\widetilde{\Lambda}_{11}', \ldots, \widetilde{\Lambda}_{1K}')$. Given $\mathbf{\Lambda}_1$, the primal variable in $(l+1)$-th iteration, e.g., $\mathbf{\Lambda}_2$, can be obtained in the same way as $\mathbf{\Lambda}_1$.

4.5.2 Update C

The process involves updating the primal variable $\mathbf{C}$. Given the updated primal variables $\mathbf{\Lambda}_1$ and $\mathbf{\Lambda}_2$ in the current iteration, $\mathbf{C}$ is obtained as

$$\mathbf{C} = \operatorname*{argmin}_{\mathbf{C}} \mathcal{L}\left(\mathbf{\Lambda}_1, \mathbf{\Lambda}_2, \mathbf{C}', \mathbf{Z}'\right) \tag{4.48}$$

$$= \operatorname*{argmin}_{\mathbf{C}} \left\{ \frac{\rho}{2} \left\| \mathbf{C}' - \left(\frac{1}{\rho}(\mathbf{Z}')^T + \mathbf{V}_1\mathbf{\Lambda}_1'\mathbf{V}_1^T + \mathbf{V}_2\mathbf{\Lambda}_2'\mathbf{V}_2^T \right) \right\|_F^2 \right\}.$$

Consider the constraints $\mathbf{I} \odot \mathbf{C} \geq \mathbf{0}$ and $\mathbf{A} \odot \mathbf{C} \leq \mathbf{0}$ in (4.37); the diagonal elements are non-negative and the off-diagonal elements are non-positive. The solution to (4.48) is obtained:

$$\mathbf{C} = \mathbf{I} \odot \left[\frac{1}{\rho}(\mathbf{Z}')^T + \mathbf{V}_1\mathbf{\Lambda}_1\mathbf{V}_1^T + \mathbf{V}_2\mathbf{\Lambda}_2\mathbf{V}_2^T \right]_+$$

$$+ \mathbf{A} \odot \left[\frac{1}{\rho}(\mathbf{Z}')^T + \mathbf{V}_1\mathbf{\Lambda}_1\mathbf{V}_1^T + \mathbf{V}_2\mathbf{\Lambda}_2\mathbf{V}_2^T \right]_- . \tag{4.49}$$

4.5.3 Update Z

Given the primary variables $\mathbf{\Lambda}_1$, $\mathbf{\Lambda}_2$, and $\mathbf{C}$, the dual variable $\mathbf{Z}$ can be updated by

$$\mathbf{Z} = \mathbf{Z}' + \rho\left(\mathbf{V}_1\mathbf{\Lambda}_1\mathbf{V}_1^T + \mathbf{V}_2\mathbf{\Lambda}_2\mathbf{V}_2^T - \mathbf{C}\right). \tag{4.50}$$

4.5.4 Results on Synthetic Data

Three synthetic random graphs generated independently from the RG model [37], ER model [55], and BA model [11] are considered. The RG model is generated with six connections per node. The ER graph is generated by randomly connecting nodes with a probability of 0.3. The BA model is generated with two initial nodes. Subsequently, each new node is connected by adding two edges to existing nodes based on preferential attachment. More than 100 independent random graphs are created for each of the three graph models. The results are averaged and plotted to provide a comprehensive evaluation.

Once a graph is generated using the three aforementioned models, the ground-truth graph Laplacian, denoted by $\mathbf{L}_0$, is derived. Subsequently, the SVD of $\mathbf{L}_0$ is performed to acquire the actual GFT, represented by $\mathbf{U}_0$. The next step involves generating the band-limited graph signals $\mathbf{Y} = \mathbf{U}_0\mathbf{S}_0 + \mathbf{\Gamma}$ with the frequency-domain signals $\mathbf{S}_0 = [\mathbf{s}_{0,1}, \cdots, \mathbf{s}_{0,M}] \in \mathbb{R}^{N \times M}$ and the additive noises $\mathbf{\Gamma} = [\boldsymbol{\gamma}_1, \cdots, \boldsymbol{\gamma}_M] \in \mathbb{R}^{N \times M}$ by following [91]. Specifically, $\mathbf{S}_0 \sim \mathcal{N}(0, \mathbf{\Lambda}_K^\dagger)$, where $\operatorname{diag}(\mathbf{\Lambda}_K) = (\lambda_1, \cdots, \lambda_K, 0, \cdots, 0)$. Two cases are considered here: $K = 3$ and $K = 15$. $\gamma_m \sim \mathcal{N}(0, \sigma^2 \mathbf{I}_N)$ is the zero-mean multivariate Gaussian noise and $\sigma^2 = 0.1$ by default.

Tables 4.1 and 4.2 provide the comparison studies of Algorithm 1 and the existing techniques with respect to the synthetic graph models. Every random graph comprises $N = 30$ nodes and observes $M = 300$ graph signal vectors, and $\beta = 0.8$. In Table 4.1, $K = 3$. In Table 4.2, $K = 15$. The knowledge of K is assumed available to the algorithm (if needed). The regularization parameter α is optimized by trying and testing for each of the considered algorithms.

The tables demonstrate that Algorithm 1 consistently outperforms all other considered techniques, achieving superior scores in both F-measure and $\rho_{\mathbf{W}}$ across all graph models. Notably, in Table 4.1, Algorithm 1 demonstrates at least a 20% improvement in F-measure relative to Sardellitti's TV and ESA methods under the ER model.

A close comparison of Tables 4.1 and 4.2 reveals that Algorithm 1 exhibits increasing superiority over the benchmark techniques as K increases from 3 to 15. This indicates that the algorithm is responsive to the signal bandwidth K and improves its performance with the rise of K.

In Tables 4.3 and 4.4, the ground truth of K are $K = 3$ and $K = 15$, respectively. The knowledge of K is unavailable to the considered algorithms here (as opposed to Tables 4.1 and 4.2). The proposed Algorithm 1 is run and its benchmarks by testing different K values (i.e., from 3 to 30 with a step size of 3). It shows that when K is consistent with its ground truth (i.e., $K = 3$ or 15 in Tables 4.3 or 4.4, respectively), Algorithm 1 can achieve the largest $\rho_{\mathbf{W}}$ or, in other words, the most accurate recovery of the graphs.

Fig. 4.1 compares the running time between the proposed algorithm (i.e., Algorithm 1) and its benchmarks, as M increases, where $K = 15$, $N = 30$, and the RG model is used to generate the graphs. It is seen that Kalofolias's model [86] is the fastest, followed by the Two-phase model [143], Dong's model [44], and Algorithm 1. Nevertheless, Algorithm 1 can achieve the highest accuracy, as shown in Table 4.4. In other words, the algorithm offers the highest prediction accuracy while maintaining a significantly low running time.

Fig. 4.2 shows the weighted adjacency matrix of the graphs learned by Algorithm 1 under different values of the coefficient β, where the RG model and ER model are considered with $K = 15$, $N = 30$, and $M = 300$. It can be seen that the settings of $\beta = 0.5$ and $\beta = 1$ allow Algorithm 1 to achieve better results than the setting of larger β values. No obvious difference can be observed between the learned graphs under the settings of $\beta = 0.5$ and $\beta = 1$. For this reason, $\beta = 0.8$ is set in the rest of this paper.

Table 4.1: Comparison of the algorithms under consideration, incl. Algorithm 1, Dong [44], Kalofolias [86], Sar-TV[140], Sar-ESA[140], Humbert [80], and Shan[143]. Here, $K = 3$, $N = 30$, and $M = 300$.

	[44]	[86]	[140]	[140]	[80]	[143]	Algo.1
RG							
F-measure	0.5154 (±0.022)	0.5229 (±0.047)	0.5341 (±0.031)	0.5084 (±0.046)	0.5193 (±0.029)	0.5468 (±0.015)	**0.5345 (±0.052)**
Recall	0.4201 (±0.023)	0.4179 (±0.026)	**0.4525 (±0.019)**	0.4017 (±0.035)	0.4422 (±0.056)	0.4325 (±0.012)	0.4221 (±0.017)
Precision	0.6667 (±0.013)	0.6984 (±0.024)	0.6516 (±0.031)	0.6290 (±0.018)	0.7441 (±0.025)	0.7432 (±0.017)	**0.7285 (±0.033)**
ρw	0.6119 (±0.022)	0.5041 (±0.040)	0.5795 (±0.025)	0.5429 (±0.023)	0.5942 (±0.017)	0.6201 (±0.011)	**0.6213 (±0.035)**
Error	0.0187 (±0.021)	**0.0156 (±0.016)**	0.0203 (±0.011)	0.0179 (±0.013)	0.0157 (±0.014)	0.0177 (±0.019)	0.0166 (±0.012)
ER							
F-measure	0.4517 (±0.021)	0.4729 (±0.036)	0.3655 (±0.022)	0.3423 (±0.019)	0.3487 (±0.024)	**0.5153 (±0.029)**	0.4956 (±0.043)
Recall	**0.4897 (±0.022)**	0.4842 (±0.013)	0.3161 (±0.016)	0.3101 (±0.026)	0.3345 (±0.031)	0.3987 (±0.027)	0.4159 (±0.026)
Precision	0.4192 (±0.018)	0.4621 (±0.016)	0.4380 (±0.023)	0.3820 (±0.017)	0.3642 (±0.022)	**0.7283 (±0.018)**	0.6131 (±0.026)
ρw	0.5501 (±0.017)	0.5656 (±0.019)	0.4373 (±0.023)	0.4417 (±0.034)	0.5255 (±0.023)	0.5905 (±0.020)	**0.5962 (±0.028)**
Error	0.0159 (±0.010)	0.0187 (±0.009)	0.0169 (±0.011)	0.0187 (±0.008)	0.0178 (±0.010)	0.0174 (±0.019)	**0.0155 (±0.008)**
BA							
F-measure	**0.5013 (±0.023)**	0.5001 (±0.016)	0.3881 (±0.025)	0.3559 (±0.031)	0.3697 (±0.033)	0.4933 (±0.023)	0.4915 (±0.020)
Recall	0.3781 (± 0.028)	**0.3966 (± 0.019)**	0.3617 (±0.013)	0.3057 (±0.033)	0.3153 (±0.029)	0.4099 (±0.015)	0.3432 (± 0.022)
Precision	0.7436 (±0.034)	0.6767 (±0.026)	0.4187 (±0.027)	0.4258 (±0.021)	0.4468 (±0.033)	0.6193 (±0.013)	**0.8649 (±0.019)**
ρw	0.4281 (±0.017)	0.4388 (±0.035)	0.4112 (±0.027)	0.3001 (±0.015)	0.4012 (±0.023)	0.5005 (±0.017)	**0.5047 (±0.027)**
Error	**0.0162 (±0.011)**	0.0173 (±0.009)	0.0197 (±0.013)	0.0199 (±0.011)	0.0212 (±0.007)	0.0197 (±0.012)	0.0231 (±0.015)

Table 4.2: Comparison of the algorithms under consideration, including Algorithm 1, Dong [44], Kalofolias [86], Sar-TV[140], Sar-ESA[140], Humbert [80], and Shan[143]. Here, $K = 15$, $N = 30$, and $M = 300$.

	[44]	[86]	[140]	[140]	[80]	[143]	Algo. 1
RG							
F-measure	0.8323 (±0.021)	0.8901 (±0.036)	0.7759 (±0.019)	0.6441 (±0.025)	0.8984 (±0.018)	0.9107 (±0.019)	**0.9228 (±0.027)**
Recall	0.7418 (±0.022)	0.8339 (±0.018)	0.6432 (±0.023)	0.4787 (±0.021)	0.8360 (±0.019)	0.8445 (±0.021)	**0.8648 (±0.031)**
Precision	0.9479 (±0.018)	0.9544 (±0.032)	0.9776 (±0.009)	0.9841 (±0.001)	0.9709 (±0.005)	0.9881 (±0.001)	**0.9891 (±0.001)**
ρw	0.8816 (±0.017)	0.8913 (±0.013)	0.8891 (±0.0171)	0.8922 (±0.016)	0.9012 (±0.021)	0.9207 (±0.012)	**0.9329 (±0.019)**
Error	0.0091 (±0.001)	0.0082 (±0.002)	0.0197 (±0.007)	0.0209 (±0.006)	0.0156 (±0.003)	**0.0135 (±0.019)**	0.0173 (±0.002)
ER							
F-measure	0.8341 (±0.026)	0.8446 (±0.019)	0.5751 (±0.031)	0.8232 (±0.028)	0.8165 (±0.016)	0.9004 (±0.018)	**0.9187 (±0.031)**
Recall	0.7771 (±0.019)	0.7871 (±0.022)	0.4305 (±0.025)	0.4253 (±0.033)	0.7510 (±0.018)	0.8900 (±0.031)	**0.864 (±0.036)**
Precision	0.9001(±0.012)	0.9112 (±0.021)	0.8660 (±0.029)	0.8879 (±0.031)	0.8945 (±0.015)	0.9110 (±0.022)	**0.9233 (±0.034)**
ρw	0.8566 (±0.018)	0.8711 (±0.021)	0.8719 (±0.025)	0.8821 (±0.022)	0.8913 (±0.011)	0.9105 (±0.015)	**0.9252 (±0.017)**
Error	0.0084 (±0.002)	0.0071 (±0.004)	0.0165 (±0.003)	0.0193 (±0.008)	0.0139 (±0.004)	0.0123 (±0.011)	**0.0119 (±0.002)**
BA							
F-measure	0.7017 (±0.021)	0.7565 (±0.029)	0.5248 (±0.037)	0.4197 (±0.033)	0.7667 (±0.017)	0.8569 (±0.033)	**0.8891 (±0.021)**
Recall	0.5909 (±0.021)	0.6568 (±0.031)	0.3812 (±0.019)	0.2788 (±0.017)	0.6677 (±0.025)	0.8079 (±0.025)	**0.8516 (±0.033)**
Precision	0.8636 (±0.022)	0.8919 (±0.031)	0.8422 (±0.019)	0.8485 (±0.015)	0.9001 (±0.024)	0.9122 (±0.011)	**0.9301 (±0.019)**
ρw	0.7090 (±0.022)	0.7276 (±0.012)	0.7966 (±0.019)	0.7872 (±0.018)	0.8869 (±0.027)	0.9099 (±0.019)	**0.9242 (±0.018)**
Error	0.0124 (±0.008)	**0.0113 (±0.007)**	0.0173 (±0.013)	0.0199 (±0.011)	0.0183 (±0.009)	0.0131 (±0.021)	0.0144 (±0.012)

Table 4.3: Comparison with different K values for considered algorithms when the ground truth is $K = 3$ under the RG model, where $N = 30$, $M = 300$, and $\beta = 0.8$.

K(adopted)	3	6	9	12	15	18	21	24	27	30
Dong [44]	0.5780 (±0.022)	–	–	–	–	–	–	–	–	–
Kalof [86]	0.5041 (±0.040)	–	–	–	–	–	–	–	–	–
Sar-TV [140]	0.5795 (±0.025)	0.5685 (±0.041)	0.5597 (±0.028)	0.5563 (±0.034)	0.5548 (±0.027)	0.5528 (±0.019)	0.5477 (±0.031)	0.5404 (±0.022)	0.5335 (±0.019)	0.5310 (±0.022)
Sar-ESA [140]	0.5429 (±0.023)	0.5347 (±0.036)	0.5301 (±0.041)	0.5296 (±0.032)	0.5265 (±0.028)	0.5252 (±0.021)	0.5185 (±0.013)	0.5119 (±0.019)	0.5054 (±0.037)	0.5001 (±0.025)
Humbert [80]	0.5942 (±0.032)	0.5893 (±0.028)	0.5814 (±0.032)	0.5777 (±0.022)	0.5703 (±0.025)	0.5645 (±0.019)	0.5516 (±0.025)	0.5598 (±0.037)	0.5575 (±0.043)	0.5531 (±0.026)
Two-phase [143]	0.6201 (±0.031)	0.6182 (±0.034)	0.6165 (±0.029)	0.6100 (±0.021)	0.6157 (±0.026)	0.6113 (±0.029)	0.6073 (±0.018)	0.6014 (±0.019)	0.59025 (±0.021)	0.5903 (±0.012)
Algorithm 1	0.6213 (±0.035)	0.6201 (±0.021)	0.6177 (±0.025)	0.6154 (±0.018)	0.6125 (±0.025)	0.6112 (±0.023)	0.6107 (±0.019)	0.6081 (±0.021)	0.6059 (±0.023)	0.6021 (±0.024)

Table 4.4: Comparison with different K values for considered algorithms when the ground truth is $K = 15$ under the RG model, where $N = 30$, $M = 300$, and $\beta = 0.8$.

K (adopted)	3	6	9	12	15	18	21	24	27	30
Dong [44]	–	–	–	–	0.8816 (±0.017)	–	–	–	–	–
Kalof [9]	–	–	–	–	0.8913 (±0.013)	–	–	–	–	–
Sar-TV [140]	0.8525 (±0.011)	0.8608 (±0.023)	0.8743 (±0.031)	0.8809 (±0.026)	**0.8891** (±0.014)	0.8807 (±0.027)	0.8850 (±0.016)	0.8835 (±0.021)	0.8822 (±0.023)	0.8819 (±0.015)
Sar-ESA [140]	0.8616 (±0.016)	0.8725 (±0.037)	0.8817 (±0.021)	0.9854 (±0.019)	**0.8922** (±0.022)	0.8910 (±0.011)	0.9899 (±0.023)	0.8868 (±0.015)	0.9353 (±0.018)	0.8796 (±0.014)
Humbert [80]	0.8603 (±0.025)	0.8685 (±0.037)	0.8917 (±0.016)	0.8989 (±0.023)	**0.9012** (±0.028)	0.9000 (±0.019)	0.8948 (±0.015)	0.8901 (±0.027)	0.8956 (±0.023)	0.8922 (±0.016)
Two-phase [143]	0.8892 (±0.012)	0.8937 (±0.024)	0.9057 (±0.019)	0.9114 (±0.011)	**0.9207** (±0.021)	0.9198 (±0.009)	0.9101 (±0.008)	0.9014 (±0.019)	0.8952 (±0.031)	0.8933 (±0.022)
Algorithm 1	0.8961 (±0.012)	0.9076 (±0.031)	0.9195 (±0.025)	0.9227 (±0.021)	**0.9329** (±0.005)	0.9295 (±0.033)	0.9273 (±0.015)	0.9287 (±0.011)	0.9252 (±0.013)	0.9249 (±0.014)

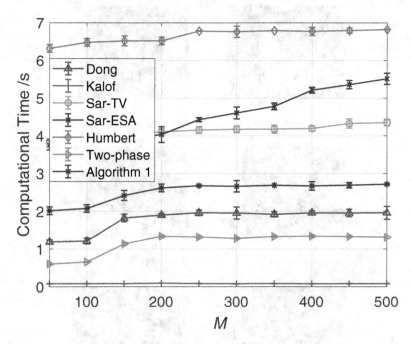

FIGURE 4.1: The running time vs. the number of M using the RG model, when $K = 15$ and $N = 30$.

4.6   Conclusion

This chapter developed a new AO-based graph learning technique to learn the graph topology of band-limited signals. It first revealed that the frequency-domain representation of the band-limited signals is a function of the graph Laplacian, thereby transforming the learning problem to only learn the graph Laplacian by estimating its eigenvectors and eigenvalues in an alternating manner. By unveiling that the feasible solution of the eigenvectors is on a Stiefel manifold, this chapter proposed to solve the eigenvectors using Stiefel manifold dual gradient descent and the eigenvalues using the ADMM.

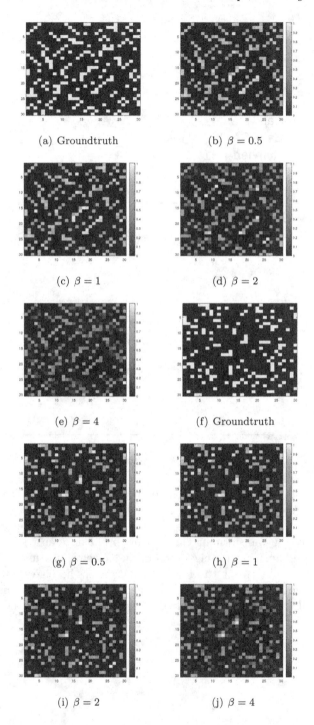

FIGURE 4.2: The learned graphs with different coefficients β under the RG and ER models, with $K = 15$ and $N = 30$.

5

Graph Learning from Band-Limited Data by Graph Fourier Transform Analysis

A graph provides an effective means to represent the statistical dependence or similarity among signals observed at different vertices. A critical challenge is to excavate graphs underlying observed signals, because of non-convex problem structure and associated high computational requirements. This chapter presents a new graph learning technique that is able to efficiently infer the graph structure underlying observed graph signals. The key idea is to reveal the intrinsic relationship between the frequency-domain representation of general band-limited graph signals and the GFT basis. Consequently, a new closed-form analytic expression for the GFT basis is derived, which depends deterministically on the observed signals, rather than being solved numerically and approximately as in the existing literature. Given the GFT basis, the estimation of the graph Laplacian, more explicitly, its eigenvalues, is convex and efficiently solved using the ADMM. Simulations based on synthetic data and experiments based on public weather datasets show that the new technique outperforms the state-of-the-art in accuracy and efficiency.

5.1 Introduction

Graph learning is the technique used to uncover the latent graph topologies of data [45]. Existing graph learning methods cannot accurately and efficiently infer the graph topology (i.e., graph Laplacian [177]) of band-limited graph signals due to difficulties in joint estimation of both the frequency-domain representation and the GFT basis converting captured data to the frequency domain. In earlier studies, underlying graph topologies were assumed to be known [154, 150]. Graph learning was used to reconstruct the graphs' underlying datasets. An overview of graph learning methods from statistical approaches to GSP-based network inference was provided in [116]. Classical graph topology inference methods, such as graph lasso [60] and covariance selection [40], estimated the covariance matrices of graph signals. More recent graph learning techniques enforced smoothness to graph signals prior to topology inference [44, 86, 62, 135, 28, 142, 131]. In other words, it has been typically assumed that the frequency-domain representations of graph signals have unlimited bandwidths, e.g., for mathematical tractability. The more general band-limited graph signals were studied in [140], where a two-step strategy approximately estimated an orthonormal sparsifying transform with AO and then recovered the graph Laplacian matrix with convex optimization. However, the fidelity of the topologies inferred from the band-limited graph signals was penalized due to the use of the AO-based approximation. In [87], a different problem was considered to learn the subspace clustering of graph signals by learning a similarity matrix, as opposed to the graph Laplacian. The AO was used to estimate the similarity matrix and a cluster/label matrix in an alternating manner.

5.2 System Model

The considered network comprises N nodes, at which a vector of signals, denoted by $\mathbf{y}_m \in \mathbb{R}^{N \times 1}$, are observed at the m-th sample $(m = 1, \cdots, M)$. M is the observed sample size. Let $\mathbf{Y} = [\mathbf{y}_1, \ldots, \mathbf{y}_M] \in \mathbb{R}^{N \times M}$ collect M observations of the band-limited signals of the network. As considered in [154] and [140], the aim is to infer the graph topology from $\mathbf{Y}$. Specifically, the goal is to characterize the network $\mathcal{G}(\mathcal{V}, \mathcal{E})$ captured by a weighted and undirected adjacency matrix $\mathbf{W} \in \mathbb{R}^{N \times N}$ with N vertices. $\mathcal{V} = \{1, \cdots, N\}$ is the set of N vertices, and $\mathcal{E} \subseteq \mathcal{V} \times \mathcal{V}$ is the set of edges. The weighted adjacency matrix $\mathbf{W}$ collects all the edges with $W_{ij} = W_{ji} \neq 0 \ \forall (i, j) \in \mathcal{E}$.

Let $\mathbf{D} \triangleq \text{diag}(\mathbf{W1})$ define the degree matrix containing the node degrees at its diagonal. Also, suppose that each node is connected to at least one other node, ensuring no isolated nodes are in the graph. In other words, none of the diagonal elements is zero in $\mathbf{D}$. Then, according to [52], the combinatorial graph Laplacian of $\mathcal{G}$ is defined as

$$\mathbf{L} = \mathbf{D} - \mathbf{W}. \tag{5.1}$$

The graph Laplacian is a semi-definite matrix with positive elements along its main diagonal and non-positive elements anywhere else [52]. By eigenvalue decomposition, $\mathbf{L}$ is rewritten as:

$$\mathbf{L} = \mathbf{U} \mathbf{\Lambda} \mathbf{U}^T, \tag{5.2}$$

where $\mathbf{\Lambda} = \text{diag}(\lambda_1, \cdots, \lambda_N)$ is the diagonal matrix containing the Laplacian eigenvalues. $\mathbf{U} = [\mathbf{u}_1, \cdots, \mathbf{u}_N]$ is an orthonormal matrix collecting all eigenvectors.

To infer the topological knowledge of $\mathcal{G}$, the Laplacian matrix $\mathbf{L}$ needs to be estimated. As done in [154], [140], and [91], $\mathbf{Y}$ is enforced to be band-limited over graph $\mathcal{G}$, e.g., the observation signals are sparse in the canonical domain [154, 140, 91]. GFT [23] has been utilized to decompose $\mathbf{Y}$ into orthonormal components $\mathbf{U}$ in the Laplacian $\mathbf{L}$ [52]. For any $m \in \{1, \cdots, M\}$, the GFT of the observed signal $\mathbf{y}_m$, denoted by $\mathbf{s}_m$, projects $\mathbf{y}_m$ onto the subspace spanned by $\mathbf{U}$, i.e., $\mathbf{s}_m = \mathbf{U}^T \mathbf{y}_m$. With the band-limited property of $\mathbf{y}_m$, $\mathbf{s}_m$ is a sparse vector and captures the key characteristics of $\mathbf{y}_m$ in the frequency domain. The band-limited signal is written as $\mathbf{y}_m = \mathbf{U}\mathbf{s}_m$. Let $\mathbf{S} = [\mathbf{s}_1, \cdots, \mathbf{s}_M] \in \mathbb{R}^{N \times M}$ collect all $\mathbf{s}_m, m = 1, \cdots, M$. It has

$$\mathbf{Y} = \mathbf{U}\mathbf{S}. \tag{5.3}$$

The frequency-domain representation $\mathbf{S}$ depends on both $\mathbf{Y}$ and the graph topology, or more explicitly, the GFT basis, $\mathbf{U}$. It is not straightforward to determine $\mathbf{S}$, given $\mathbf{Y}$. $\mathbf{S}$ and $\mathbf{U}$ need to be jointly estimated, as done in the existing literature [140]. With the sparsity of $\mathbf{s}_m$ ($m \in \{1, \cdots, M\}$), $\mathbf{S} \in \mathcal{B}_K$ is set as a K-block sparse matrix with multiple all-zero row-vectors. K specifies the bandwidth of the frequency-domain representation of the observed band-limited graph signal $\mathbf{Y}$. $\mathcal{B}_K$ collects all K-block sparse matrices [23]:

$$\mathcal{B}_K \triangleq \{\mathbf{S} \in \mathbb{R}^{N \times M}, \mathbf{S}(i,:) = \mathbf{0}, i \notin \mathcal{K} \subseteq \mathcal{V}, K = |\mathcal{K}|\}, \tag{5.4}$$

where $\mathbf{S}(i,:)$ is the i-th row of $\mathbf{S}$, and $\mathcal{K} \in \mathcal{V}$.

5.3 Problem Statement

The Laplacian $\mathbf{L}$ is estimated and, in turn, the topological knowledge of graph $\mathcal{G}$ substantiates the observation $\mathbf{Y}$. Given the band-limitedness of $\mathbf{Y}$, the problem is casted as

$$\min_{\mathbf{L}, \mathbf{U} \in \mathbb{R}^{N \times N}, \mathbf{S} \in \mathbb{R}^{N \times M}} \|\mathbf{Y} - \mathbf{U}\mathbf{S}\|_F^2 + f(\mathbf{L}, \mathbf{Y}) \tag{5.5a}$$

$$\text{s.t.} \quad \mathbf{U}\mathbf{U}^T = \mathbf{I}_N, \tag{5.5b}$$

$$\mathbf{S} \in \mathcal{B}_K, \tag{5.5c}$$

$$\mathbf{L} = \mathbf{U}\mathbf{\Lambda}\mathbf{U}^T, \mathbf{L} \in \mathbb{L}, \tag{5.5d}$$

$$\mathbf{u}_1 = \frac{1}{\sqrt{N}}\mathbf{1}, \tag{5.5e}$$

where $\|\cdot\|_F$ stands for the Frobenius norm. The objective (5.5a) consists of two terms. The first term of (5.5a) penalizes any discrepancy between $\mathbf{U}\mathbf{S}$ and $\mathbf{Y}$. The second term characterizes a regularized function [44, 140]. The details of $f(\mathbf{L}, \mathbf{Y})$ are provided in Section 5.5. Constraint (5.5b) ensures $\mathbf{U}$ to be a unitary matrix complying with (5.2). Constraint (5.5c) imposes the K-block sparsity of $\mathbf{S}$ in (5.4). Constraint (5.5d) guarantees that $\mathbf{L}$ refers to a valid Laplacian matrix that satisfies the necessary conditions and properties, and $\mathbb{L}$ collects all valid candidates to $\mathbf{L}$ [52], i.e.,

$$\mathbb{L} = \{\mathbf{L} \succeq \mathbf{0} | \mathbf{L}\mathbf{1} = \mathbf{0}, L_{ij} \leq 0, i \neq j\}, \tag{5.6}$$

where $\mathbf{0}$ stands for an all-zero vector. Since $\mathbf{L1} = \mathbf{0}$ in (5.6), 0 is an eigenvalue of $\mathbf{L}$, leading to $\mathbf{u}_1 = \frac{1}{\sqrt{N}}\mathbf{1}$ in (5.5e). $\mathbf{u}_1$ is the first column of $\mathbf{U}$.

Remark 1 *Problem (5.5) is non-convex due to the orthonormal property in (5.5b) and the sparse property in (5.5c). (5.5) is decoupled into two phases. The first phase is that, given the observation* $\mathbf{Y}$*,* $\mathbf{U}$ *is estimated, as described in Section 5.4. Based on the estimated* $\mathbf{U}$*, the second phase estimates the eigenvalues of* $\mathbf{L}$*,* $\mathbf{\Lambda}$*, as delineated in Section 5.5.*

5.4   Closed-Form Expression for GFT Basis

Given the observed signal $\mathbf{Y}$, problem (5.5) is solved by first estimating the GFT basis $\mathbf{U}$ to minimize $\|\mathbf{Y} - \mathbf{US}\|_F^2$ subject to $\mathbf{UU}^T = \mathbf{I}_N$, $\mathbf{S} \in \mathcal{B}_K$, and $\mathbf{u}_1 = \frac{1}{\sqrt{N}}\mathbf{1}$ (and then estimating the eigenvalues $\mathbf{\Lambda}$ to minimize the regularizer $f(\mathbf{Y}, \mathbf{L})$ given the obtained $\mathbf{U}$, as will be discussed in Section 5.5). The reason for beginning with the GFT basis, $\mathbf{U}$, is the paramount importance of $\mathbf{U}$ in graph learning and of the capability to recover $\mathbf{U}$ from $\mathbf{Y}$ in the absence of the *a-priori* knowledge about the statistical distribution of $\mathbf{L}$.

According to the definition $\mathbf{Y} = \mathbf{US}$ and the orthonormality of the unitary matrix $\mathbf{U}$ in (5.5b), [140, eq. (8)] is first solved

$$\min_{\mathbf{U}\in\mathbb{R}^{N\times N}, \mathbf{S}\in\mathbb{R}^{N\times M}} \|\mathbf{Y} - \mathbf{US}\|_F^2 = \|\mathbf{U}^T\mathbf{Y} - \mathbf{S}\|_F^2, \text{ s.t. (5.5b), (5.5c), (5.5e).}$$
(5.7)

Despite its convex objective, problem (5.7) is non-convex due to the non-convexity of (5.5b) and (5.5c). (5.7) is reorganized as

$$\min_{\mathbf{U}\in\mathbb{R}^{N\times N}, \mathbf{S}\in\mathcal{B}_K} \sum_{i=1}^{N} \|\mathbf{u}_i^T\mathbf{Y} - \mathbf{S}(i,:)\|_2^2, \text{ s.t. (5.5b), (5.5e),}$$
(5.8)

which can be rewritten as

$$\min_{\mathbf{U}\in\mathbb{R}^{N\times N}, \mathcal{K}} \left(\min_{\mathbf{S}\in\mathcal{B}_K} \sum_{i\in\mathcal{K}} \|\mathbf{u}_i^T\mathbf{Y} - \mathbf{S}(i,:)\|_2^2 + \sum_{i\notin\mathcal{K}} \|\mathbf{u}_i^T\mathbf{Y}\|_2^2 \right), \text{ s.t. (5.5b), (5.5e).}$$
(5.9)

By assessing the objective of (5.9), it is found that the optimal index set, $\mathcal{K}$, collects the indices to the K largest entries of $\{\|\mathbf{u}_i^T\mathbf{Y}\|\}_i^N$, and

$$\mathbf{S}(i,:) = \begin{cases} \mathbf{u}_i^T\mathbf{Y}, & \text{if } i \in \mathcal{K}; \\ \mathbf{0}, & \text{if } i \notin \mathcal{K}. \end{cases}$$
(5.10)

Only the $(N - K)$ smallest entries of $\{\|\mathbf{u}_i^T \mathbf{Y}\|\}_i^N$ remain in the objective of (5.9) after $\mathbf{S}$ is optimized to suppress $\sum_{i\in\mathcal{K}} \|\mathbf{u}_i^T \mathbf{Y} - \mathbf{S}(i, :)\|_2^2$ in (5.10). The objective is minimized with respect to $\mathbf{S}$. The next step is to find the optimal $\mathbf{U}$, denoted by $\mathbf{U}^*$, to minimize the objective in (5.7).

By plugging (5.10) into the objective of (5.9), (5.7) becomes

$$\mathbf{U}^* = \underset{\mathbf{U},\mathcal{K}}{\arg\min} \sum_{i\notin\mathcal{K}} \|\mathbf{u}_i^T \mathbf{Y}\|_2^2 = \underset{\mathbf{U},\mathcal{K}}{\arg\min} \|\mathbf{U}_{\mathcal{K}^c}^T \mathbf{Y}\|_F^2 = \underset{\mathbf{U},\mathcal{K}}{\arg\max} \|\mathbf{U}_{\mathcal{K}}^T \mathbf{Y}\|_F^2.$$

$$(5.11)$$

Here, $\mathcal{K}^c$ is the complementary set of $\mathcal{K}$, i.e., $\mathcal{K}^c = \mathcal{V} \setminus \mathcal{K}$. $\mathbf{U}_{\mathcal{K}}$ and $\mathbf{U}_{\mathcal{K}^c}$ are the matrices collating the column-vectors of $\mathbf{U}$ indexed by $\mathcal{K}$ and $\mathcal{K}^c$, respectively.

Despite the non-convexity of (5.11) in $\mathcal{K}$, it is noted that (5.11) is to find the K-dimensional space on which $\mathbf{Y}$ has the largest orthogonal projection:

$$\underset{\mathbf{U},\mathcal{K}}{\arg\max} \|\mathbf{U}_{\mathcal{K}}^T \mathbf{Y}\|_F^2 = \underset{\mathbf{U},\mathcal{K}}{\arg\max} \operatorname{tr}(\mathbf{Y}^T \mathbf{U}_{\mathcal{K}} \mathbf{U}_{\mathcal{K}}^T \mathbf{Y}) = \underset{\mathbf{U},\mathcal{K}}{\arg\max} \operatorname{tr}(\mathbf{P}_{\mathbf{U}_{\mathcal{K}}} \mathbf{Y}\mathbf{Y}^T), \quad (5.12)$$

where $\mathbf{P}_{\mathbf{U}_{\mathcal{K}}} = \mathbf{U}_{\mathcal{K}} \mathbf{U}_{\mathcal{K}}^T$ is the orthogonal projection on the subspace spanned by the column-vectors of $\mathbf{U}_{\mathcal{K}}$.

Lemma 1 *The use of the orthogonal projection, $\mathbf{P}_{\mathbf{U}_{\mathcal{K}}} = \mathbf{U}_{\mathcal{K}} \mathbf{U}_{\mathcal{K}}^T$, preserves the orthogonality constraint (5.5b) in (5.12).*

Proof 1 *See Appendix 10.1.*

By applying (5.12), problem (5.7) can be rewritten as

$$\mathbf{U}^* = \arg\max_{\mathbf{U},\mathcal{K}} \operatorname{tr}(\mathbf{P}_{\mathbf{U}_{\mathcal{K}}} \mathbf{Y}\mathbf{Y}^T), \quad \text{s.t. (5.5e).} \quad (5.13)$$

Theorem 1 *Considering both the cases of $\mathbf{u}_1 \notin \mathbf{U}_{\mathcal{K}}$ and $\mathbf{u}_1 \in \mathbf{U}_{\mathcal{K}}$, the closed-form optimal solution $\mathbf{U}^* = [\mathbf{U}_{\mathcal{K}}^*, \mathbf{U}_{\mathcal{K}^c}^*]$ to problem (5.7) is given by*

$$\mathbf{U}^* = \operatorname{eigen}\left[(\mathbf{I} - \mathbf{u}_1 \mathbf{u}_1^T) \mathbf{Y}\mathbf{Y}^T (\mathbf{I} - \mathbf{u}_1 \mathbf{u}_1^T)^T\right], \quad (5.14)$$

where $\operatorname{eigen}[\mathbf{X}]$ provides the eigenvectors of $\mathbf{X}$.

Proof 2 *The solution to (5.13) is derived in the two cases:*

1. *In the case of $\mathbf{u}_1 \notin \mathbf{U}_{\mathcal{K}}$: Let $\mathbf{P}_{\mathbf{U}_{\mathcal{K}\setminus\{1\}}}$ represent the orthogonal projection of the subspace of $\mathbf{U}_{\mathcal{K}}$, where $\mathbf{P}_{\mathbf{U}_{\mathcal{K}\setminus\{1\}}} = \mathbf{P}_{\mathbf{U}_{\mathcal{K}}}(\mathbf{I} - \mathbf{u}_1 \mathbf{u}_1^T)(\mathbf{I} - \mathbf{u}_1 \mathbf{u}_1^T)^T$. The objective of (5.13) is rewritten as*

$$\max_{\mathbf{U},\mathcal{K}} \operatorname{tr}(\mathbf{P}_{\mathbf{U}_{\mathcal{K}}}(\mathbf{I} - \mathbf{u}_1 \mathbf{u}_1^T) \mathbf{Y}\mathbf{Y}^T (\mathbf{I} - \mathbf{u}_1 \mathbf{u}_1^T)^T). \quad (5.15)$$

   Lemma 2 *If $\mathbf{u}_1 \notin \mathbf{U}_{\mathcal{K}}$, the solution $\mathbf{U}_{\mathcal{K}}^*$ comprises the eigenvectors corresponding to the K largest eigenvalues of $(\mathbf{I} - \mathbf{u}_1 \mathbf{u}_1^T) \mathbf{Y}\mathbf{Y}^T (\mathbf{I} - \mathbf{u}_1 \mathbf{u}_1^T)^T$.*

Proof 3 *See Appendix 10.2.*

2. In the case of $\mathbf{u}_1 \in \mathbf{U}_{\mathcal{K}}$*: By writing* $\mathbf{U}_{\mathcal{K}} = \left[\mathbf{u}_1, \mathbf{U}_{\mathcal{K} \setminus \{1\}}\right]$*, the objective of (5.13) can be rewritten as*

$$\underset{\mathbf{U},\mathcal{K}}{\arg\max} \; \mathrm{tr}\left(\mathbf{P}_{\mathbf{U}_{\mathcal{K}}} \mathbf{Y}\mathbf{Y}^{\mathrm{T}}\right) = \arg\max_{\mathbf{U},\mathcal{K}} \mathrm{tr}(\mathbf{P}_{\mathbf{U}_{\mathcal{K} \setminus \{1\}}} \mathbf{Y}\mathbf{Y}^{\mathrm{T}} + \mathbf{u}_1\mathbf{u}_1^{\mathrm{T}}\mathbf{Y}\mathbf{Y}^{\mathrm{T}})$$

$$(5.16\mathrm{a})$$

$$= \arg\max_{\mathbf{U},\mathcal{K}} \mathrm{tr}(\mathbf{P}_{\mathbf{U}_{\mathcal{K}}} \left(\mathbf{I} - \mathbf{u}_1\mathbf{u}_1^{T}\right) \mathbf{Y}\mathbf{Y}^{T} \left(\mathbf{I} - \mathbf{u}_1\mathbf{u}_1^{T}\right)^{T}), \quad (5.16\mathrm{b})$$

where $\mathbf{u}_1\mathbf{u}_1^{T}\mathbf{Y}\mathbf{Y}^{T}$ *is a constant in (5.16a) and suppressed. Apparently, (5.16b) is identical to (5.15). The solution* $\mathbf{U}_{\mathcal{K}}^{*}$ *comprises the eigenvectors corresponding to the* $(K - 1)$ *largest eigenvalues of* $\left(\mathbf{I} - \mathbf{u}_1\mathbf{u}_1^{T}\right) \mathbf{Y}\mathbf{Y}^{T} \left(\mathbf{I} - \mathbf{u}_1\mathbf{u}_1^{T}\right)^{T}$*, and* $\mathbf{u}_1$*, which can be proved in the same way as* **Lemma 2**.

Remark 2 (5.14) *is derived analytically by solving exactly the challenging non-convex problem (5.7). The computational cost of (5.14) is only* $\mathcal{O}(MN^2)$*. The closed-form solution is achieved by revealing unprecedentedly the relation between the GFT coefficient* $\mathbf{S}$ *and the GFT basis* $\mathbf{U}$*, as done in (5.10); then reformulating losslessly the joint estimation of* $\mathbf{S}$ *and* $\mathbf{U}$ *in (5.7) to the estimation of* $\mathbf{U}$ *only in (5.11); and finally solving* $\mathbf{U}^*$ *with non-trivial analysis, as shown in the proof of* **Theorem 1**. *In contrast, the existing techniques, e.g., the one developed in [140], solved problem (5.7) numerically and approximately by using the AO. Given the non-convexity of problem (5.7), the AO algorithm in [140] could only solve* $\mathbf{S}$ *and* $\mathbf{U}$ *in an alternating manner. Specifically, at the k-th iteration of the algorithm, given* $\mathbf{U}^{(k-1)}$*,* $\mathbf{S}^{(k)} = \arg\min_{\mathbf{S} \in \mathbb{R}^{N \times M}} \|(\mathbf{U}^{(k-1)})^T\mathbf{Y} - \mathbf{S}\|_F^2, \text{s.t. } \mathbf{S} \in \mathcal{B}_K$*, and then given* $\mathbf{S}^{(k)}$*,* $\mathbf{U}^{(k)}$ *is obtained by solving problem* $\mathbf{U}^{(k)} = \arg\min_{\mathbf{S} \in \mathbb{R}^{N \times M}} \|\mathbf{U}^T\mathbf{Y} - \mathbf{S}^{(k)}\|_F^2, \text{s.t. } \mathbf{U}\mathbf{U}^T = \mathbf{I}_N$*. Not only does this provide a suboptimal solution to (5.7), but incurs a high computational cost of* $\mathcal{O}(MN^2)$ *per iteration.*

Remark 3 *While the analysis of* $\mathbf{U}$ *and* $\mathbf{S}$ *depends on a given* K*, the result of the analysis, i.e.,* $\mathbf{U}^* = \mathrm{eigen}[(\mathbf{I} - \mathbf{u}_1\mathbf{u}_1^{T})\mathbf{Y}\mathbf{Y}^{T}(\mathbf{I} - \mathbf{u}_1\mathbf{u}_1^{T})^{T}]$ *in (5.14), turns out to be applicable to any* K*,* $1 \leq K \leq N - 1$*. To this end,* $K = N - 1$ *can be initially set to evaluate all frequency-domain components of the observed signal* $\mathbf{Y}$ *and analytically derive* $\mathbf{U}^*$ *and* $\mathbf{S} = (\mathbf{U}^*)^T\mathbf{Y}$*. Then, the norm of each row of* $\mathbf{S}$ *is assessed, i.e.,* $\|\mathbf{S}(i,:)\|$*. The value of* K *is determined either by counting the number of rows with their norms (or the total of their norms) exceeding a preconfigured threshold, or by sorting the norms in descending order and identifying* K *with an elbow method.*

5.5   Laplacian Eigenvalues Estimation

With the *a-priori* knowledge of statistics or distribution of $\mathbf{L}$, the regularizer $f(\mathbf{L}, \mathbf{Y})$ can capture the desired characteristics of the topology, e.g., the sparsity of $\mathbf{L}$. Given the GFT basis $\mathbf{U}^* \in \mathbb{R}^{N \times N}$ in (5.14) and the definition $\mathbf{L} = \mathbf{U}\mathbf{\Lambda}\mathbf{U}^T$, to estimate the eigenvalues of $\mathbf{L}$ is in essence to estimate the N eigenvalues, $\mathbf{\Lambda}$, as done in this section. Typically, $\mathbf{L}$ follows an exponential distribution [52]. The MAP estimate of $\mathbf{L}$ is written as [52]

$$\min_{\mathbf{L}} \ \text{tr}(\mathbf{L}\mathbf{Y}\mathbf{Y}^T) - M \log|\mathbf{L}| + \alpha \|\text{vec}(\mathbf{L})\|_1, \ \text{s.t.} \ \mathbf{L} = \mathbf{U}\mathbf{\Lambda}\mathbf{U}^T, \mathbf{L} \in \mathbb{L}. \quad (5.17)$$

In this sense, $f(\mathbf{L}, \mathbf{Y}) = \text{tr}(\mathbf{L}\mathbf{Y}\mathbf{Y}^T) - \log|\mathbf{L}| + \alpha \|\text{vec}(\mathbf{L})\|_1$ provides a reasonable regularizer for (5.5a). $\|\cdot\|_1$ stands for ℓ_1-norm, and $|\cdot|$ denotes pseudo-determinant. α is a tunable regularization parameter. With the non-positivity of $\mathbf{L}$ and $\mathbf{L}\mathbf{1} = 0$, $\|\text{vec}(\mathbf{L})\|_1 = 2\,\text{tr}(\mathbf{L})$. By using the linearity of trace, $f(\mathbf{L}, \mathbf{Y})$ is rewritten as

$$f(\mathbf{L}, \mathbf{Y}) = \text{tr}\left(\mathbf{L}\mathbf{Y}\mathbf{Y}^T + 2\alpha\mathbf{L}\right) - M \log|\mathbf{L}| = \text{tr}(\mathbf{L}\mathbf{T}) - M \log|\mathbf{L}|, \quad (5.18)$$

where $\mathbf{T} = \mathbf{Y}\mathbf{Y}^T + 2\alpha\mathbf{I}$. It is noted that $f(\mathbf{L}, \mathbf{Y}) = \text{tr}(\mathbf{L}\mathbf{T}) - M \log|\mathbf{L}|$ is still intractable due to the singular pseudo-determinant of $|\mathbf{L}|$. To circumvent this impasse, the equivalence of $\text{tr}(\mathbf{L}\mathbf{T}) - M \log|\mathbf{L}|$ and $\text{tr}(\mathbf{L}\mathbf{T}) - M \log\det(\mathbf{L} + \mathbf{J})$ with $\mathbf{J} = \mathbf{u}_1\mathbf{u}_1^T = \frac{1}{N}\mathbf{1}\mathbf{1}^T$ [52, Prop. 1] is explored, and rewritten (5.17) as

$$\min_{\mathbf{L}} \text{tr}(\mathbf{L}\mathbf{T}) - M \log\det(\mathbf{L} + \mathbf{J})$$
$$\text{s.t.} \ \mathbf{L} = \mathbf{U}\mathbf{\Lambda}\mathbf{U}^T, \ \mathbf{L} \succeq 0, \mathbf{L}\mathbf{1} = 0, \ \mathbf{I} \odot \mathbf{L} \geq 0, \ \mathbf{A} \odot \mathbf{L} \leq 0, \quad (5.19)$$

where the constraint is from (5.6) and $\mathbf{A} = \mathbf{1}\mathbf{1}^T - \mathbf{I}$.

With the optimal $\mathbf{U}^*$ obtained in (5.14), $\mathbf{L}$ is rewritten as in the following two cases:

1. *In the case of* $\mathbf{u}_1 \notin \mathbf{U}_{\mathcal{K}}^*$*, the eigenvectors of* $\mathbf{L}$*, i.e.,* $\mathbf{U}^*$*, is arranged as* $[\mathbf{u}_1, \mathbf{U}_{\mathcal{K}}^*, \mathbf{U}_{\mathcal{K}^c\setminus\{1\}}^*]$*. By taking eigenvalue decomposition,* $\mathbf{L}$ *is written as*

$$\mathbf{L} = \left[\mathbf{u}_1, \mathbf{U}_{\mathcal{K}}^*, \mathbf{U}_{\mathcal{K}^c\setminus\{1\}}^*\right] \mathbf{\Phi}_1 \left[\mathbf{u}_1, \mathbf{U}_{\mathcal{K}}^*, \mathbf{U}_{\mathcal{K}^c\setminus\{1\}}^*\right]^T \quad (5.20a)$$

$$= \left[\mathbf{U}_{\mathcal{K}}^*, \mathbf{U}_{\mathcal{K}^c\setminus\{1\}}^*\right] \mathbf{\Phi}_2 \left[\mathbf{U}_{\mathcal{K}}^*, \mathbf{U}_{\mathcal{K}^c\setminus\{1\}}^*\right]^T, \quad (5.20b)$$

where $\mathbf{\Phi}_1 = \text{blkdiag}\left(0, \mathbf{\Lambda}_{\mathcal{K}}, \mathbf{\Lambda}_{\mathcal{K}^c\setminus\{1\}}\right)$ *is a block-diagonal matrix with* $\mathbf{\Lambda}_{\mathcal{K}} \succeq 0$ *and* $\mathbf{\Lambda}_{\mathcal{K}} \in \mathbb{R}^{K \times K}$ *and collects all the eigenvalues of* $\mathbf{L}$ *associated with the eigenvectors* $[\mathbf{u}_1, \mathbf{U}_{\mathcal{K}}^*, \mathbf{U}_{\mathcal{K}^c\setminus\{1\}}^*]$*; and* $\mathbf{\Phi}_2 = \text{blkdiag}\left(\mathbf{\Lambda}_{\mathcal{K}}, \mathbf{\Lambda}_{\mathcal{K}^c\setminus\{1\}}\right)$ *with* $\mathbf{\Lambda}_{\mathcal{K}^c\setminus\{1\}} \succeq 0$ *and* $\mathbf{\Lambda}_{\mathcal{K}^c\setminus\{1\}} \in \mathbb{R}^{(N-K-1)\times(N-K-1)}$*.*

2. *In the case of* $\mathbf{u}_1 \in \mathbf{U}_{\mathcal{K}}^*$*,* $\mathbf{U}^*$ *is arranged as* $[\mathbf{u}_1, \mathbf{U}_{\mathcal{K}\setminus\{1\}}^*, \mathbf{U}_{\mathcal{K}^c}^*]$ *and* $\mathbf{L}$ *is*

$$\mathbf{L} = \left[\mathbf{u}_1, \mathbf{U}_{\mathcal{K}\setminus\{1\}}^*, \mathbf{U}_{\mathcal{K}^c}^*\right] \mathbf{\Phi}_1 \left[\mathbf{u}_1, \mathbf{U}_{\mathcal{K}\setminus\{1\}}^*, \mathbf{U}_{\mathcal{K}^c}^*\right]^T \quad (5.21a)$$

$$= \left[\mathbf{U}_{\mathcal{K}\setminus\{1\}}^*, \mathbf{U}_{\mathcal{K}^c}^*\right] \mathbf{\Phi}_2 \left[\mathbf{U}_{\mathcal{K}\setminus\{1\}}^*, \mathbf{U}_{\mathcal{K}^c}^*\right]^T, \quad (5.21b)$$

where $\mathbf{\Phi}_1 = \text{blkdiag}\left(0, \mathbf{\Lambda}_{\mathcal{K}\backslash\{1\}}, \mathbf{\Lambda}_{\mathcal{K}^c}\right)$ with $\mathbf{\Lambda}_{\mathcal{K}\backslash\{1\}} \succeq \mathbf{0}$ and $\mathbf{\Lambda}_{\mathcal{K}\backslash\{1\}} \in \mathbb{R}^{(K-1)\times(K-1)}$ collects all the eigenvalues of $\mathbf{L}$ associated with the eigenvectors $[\mathbf{u}_1, \mathbf{U}^*_{\mathcal{K}^c\backslash\{1\}}, \mathbf{U}^*_{\mathcal{K}}]$; and $\mathbf{\Phi}_2 = \text{blkdiag}\left(\mathbf{\Lambda}_{\mathcal{K}\backslash\{1\}}, \mathbf{\Lambda}_{\mathcal{K}^c}\right)$ with $\mathbf{\Lambda}_{\mathcal{K}^c} \succeq \mathbf{0}$ and $\mathbf{\Lambda}_{\mathcal{K}^c} \in \mathbb{R}^{(N-K)\times(N-K)}$.

To unify the presentation of the two cases, the following definitions are provided: 1) If $\mathbf{u}_1 \notin \mathbf{U}^*_{\mathcal{K}}$, then $\mathbf{V}_1 = \mathbf{U}^*_{\mathcal{K}}$, $\mathbf{V}_2 = \mathbf{U}^*_{\mathcal{K}^c\backslash\{1\}}$, $\mathbf{\Lambda}_1 = \mathbf{\Lambda}_{\mathcal{K}}$, and $\mathbf{\Lambda}_2 = \mathbf{U}_{\mathcal{K}^c\backslash\{1\}}$; 2) If $\mathbf{u}_1 \in \mathbf{U}^*_{\mathcal{K}}$, then $\mathbf{V}_1 = \mathbf{U}^*_{\mathcal{K}^c\backslash\{1\}}$, $\mathbf{V}_2 = \mathbf{U}^*_{\mathcal{K}}$, $\mathbf{\Lambda}_1 = \mathbf{\Lambda}_{\mathcal{K}^c\backslash\{1\}}$, and $\mathbf{\Lambda}_2 = \mathbf{\Lambda}_{\mathcal{K}}$. As a result, the Laplacian matrix $\mathbf{L} = \mathbf{U}\mathbf{\Lambda}\mathbf{U}^T$ in (5.19) can be written as

$$\mathbf{L} = [\mathbf{V}_1, \mathbf{V}_2]\,\mathbf{\Phi}_2\,[\mathbf{V}_1, \mathbf{V}_2]^T, \tag{5.22}$$

where $\mathbf{\Phi}_2 = \text{blkdiag}(\mathbf{\Lambda}_1, \mathbf{\Lambda}_2)$. Since $[\mathbf{V}_1, \mathbf{V}_2]^T\mathbf{u}_1 = \mathbf{0}$, $\mathbf{L}\mathbf{1} = \mathbf{0}$ in (5.19) is preserved in (5.22). By plugging (5.22), $\text{tr}(\mathbf{L}\mathbf{T})$ can be rewritten as

$$\text{tr}(\mathbf{L}\mathbf{T}) = \text{tr}(\mathbf{V}_1\mathbf{\Lambda}_1\mathbf{V}_1^T\mathbf{T}) + \text{tr}(\mathbf{V}_2\mathbf{\Lambda}_2\mathbf{V}_2^T\mathbf{T}) = \text{tr}(\mathbf{\Lambda}_1\widetilde{\mathbf{T}}_1) + \text{tr}(\mathbf{\Lambda}_2\widetilde{\mathbf{T}}_2), \tag{5.23}$$

where $\widetilde{\mathbf{T}}_1 = \mathbf{V}_1^T\mathbf{T}\mathbf{V}_1$ and $\widetilde{\mathbf{T}}_2 = \mathbf{V}_2^T\mathbf{T}\mathbf{V}_2$. By plugging (5.22) and $\mathbf{J} = \frac{1}{N}\mathbf{1}\mathbf{1}^T$, $\log\det\left(\mathbf{L} + \mathbf{J}\right)$ can be rewritten as

$$\begin{aligned}
\log\det\left(\mathbf{L} + \mathbf{J}\right) &= \log\det\left(\left[\frac{1}{\sqrt{N}}, \mathbf{V}_1, \mathbf{V}_2\right]\begin{bmatrix}1 & \\ & \mathbf{\Phi}_2\end{bmatrix}\left[\frac{1}{\sqrt{N}}, \mathbf{V}_1, \mathbf{V}_2\right]^T\right) \\
&= \log\det\left(\text{blkdiag}\left(1, \mathbf{\Lambda}_1, \mathbf{\Lambda}_2\right)\right) \\
&= \log\det\left(\mathbf{\Lambda}_1\right) + \log\det\left(\mathbf{\Lambda}_2\right).
\end{aligned} \tag{5.24}$$

Constraint $\mathbf{L}\mathbf{1} = \mathbf{0}$ in (5.19) is preserved in (5.22), and so is it in (5.23) and (5.24). The equality constraint $\mathbf{L} = \mathbf{U}\mathbf{\Lambda}\mathbf{U}^T$ in problem (5.19) is retained throughout the analysis, i.e., in (5.22) and then (5.23) and (5.24). By substituting (5.23) and (5.24), problem (5.19) becomes

$$\min_{\mathbf{\Lambda}_1, \mathbf{\Lambda}_2} \left[\text{tr}(\mathbf{\Lambda}_1\widetilde{\mathbf{T}}_1) + \text{tr}(\mathbf{\Lambda}_2\widetilde{\mathbf{T}}_2) - M\log\det\left(\mathbf{\Lambda}_1\right) - M\log\det\left(\mathbf{\Lambda}_2\right)\right] \tag{5.25a}$$

$$\text{s.t.} \quad \mathbf{\Lambda}_1 \succeq \mathbf{0}, \mathbf{\Lambda}_2 \succeq \mathbf{0}, \tag{5.25b}$$

$$\mathbf{I} \odot [\mathbf{V}_1\mathbf{\Lambda}_1\mathbf{V}_1^T + \mathbf{V}_2\mathbf{\Lambda}_2\mathbf{V}_2^T] \geq \mathbf{0}, \tag{5.25c}$$

$$\mathbf{A} \odot [\mathbf{V}_1\mathbf{\Lambda}_1\mathbf{V}_1^T + \mathbf{V}_2\mathbf{\Lambda}_2\mathbf{V}_2^T] \leq \mathbf{0}. \tag{5.25d}$$

Here, $\mathbf{\Lambda}_1$ and $\mathbf{\Lambda}_2$ may not have full rank (since $\mathbf{U}^*_{\mathcal{K}}$ and $\mathbf{U}^*_{\mathcal{K}^c}$ are organized against $(\mathbf{I} - \mathbf{u}_1\mathbf{u}_1^T)\mathbf{Y}\mathbf{Y}^T(\mathbf{I} - \mathbf{u}_1\mathbf{u}_1^T)^T$ in (5.14), not against the Laplacian matrix $\mathbf{L}$). $\mathbf{\Lambda}_1$ and $\mathbf{\Lambda}_2$ can have zero diagonal elements, if the graph $\mathcal{G}$ is not a connected graph. Problem (5.25) is not continually differentiable, and cannot be solved using CVX. By defining $\mathbf{C} = \mathbf{L}$, (5.25) is rewritten as

$$\min_{\mathbf{\Lambda}_1, \mathbf{\Lambda}_2} \left[\text{tr}(\mathbf{\Lambda}_1\widetilde{\mathbf{T}}_1) + \text{tr}(\mathbf{\Lambda}_2\widetilde{\mathbf{T}}_2) - M\log\det\left(\mathbf{\Lambda}_1\right) - M\log\det\left(\mathbf{\Lambda}_2\right)\right] \tag{5.26a}$$

$$\text{s.t.} \quad \mathbf{\Lambda}_1 \succeq \mathbf{0}, \mathbf{\Lambda}_2 \succeq \mathbf{0}, \tag{5.26b}$$

$$\mathbf{V}_1\mathbf{\Lambda}_1\mathbf{V}_1^T + \mathbf{V}_2\mathbf{\Lambda}_2\mathbf{V}_2^T - \mathbf{C} = \mathbf{0}, \tag{5.26c}$$

$$\mathbf{I} \odot \mathbf{C} \geq \mathbf{0}, \ \mathbf{A} \odot \mathbf{C} \leq \mathbf{0}, \tag{5.26d}$$

which can be solved using the ADMM with details provided in Appendix 10.3.

Convergence Analysis. The ADMM solves constrained optimization problems [163] with the following structure: $\min_{\mathbf{x},\mathbf{y}} f(\mathbf{x}) + g(\mathbf{y})$, s.t. $\mathbf{Ax} - \mathbf{By} = \mathbf{E}$, where $\mathbf{x} \in \mathbb{R}^{N \times 1}$ and $\mathbf{y} \in \mathbb{R}^{N \times 1}$ denotes the variables being optimized; $\mathbf{A} \in \mathbb{R}^{N \times N}$, $\mathbf{B} \in \mathbb{R}^{N \times N}$, and $\mathbf{E} \in \mathbb{R}^{N \times N}$ are known; $f(\cdot)$ and $g(\cdot)$ represent convex objectives. Therefore, the convergence of the ADMM is guaranteed under the following two conditions: 1) Both $f(\cdot)$ and $g(\cdot)$ are closed, proper and convex sets; hence, the subproblems arising from the updates of $\mathbf{x}$ and $\mathbf{y}$ are solvable; and 2) The Lagrangian has a saddle point. Problem (5.26) satisfies the two conditions. First, $\text{tr}(\mathbf{\Lambda}_1 \widetilde{\mathbf{T}}_1) - M \log \det (\mathbf{\Lambda}_1)$ and $\text{tr}(\mathbf{\Lambda}_2 \widetilde{\mathbf{T}}_2) - M \log \det (\mathbf{\Lambda}_2)$ in (5.26) correspond to $f(\cdot)$ and $g(\cdot)$, respectively. $\mathbf{\Lambda}_1$ and $\mathbf{\Lambda}_2$ are strictly convex with proper, closed, and convex sets. The first condition is satisfied. The primary variables $\mathbf{\Lambda}_1$ and $\mathbf{\Lambda}_2$ are solvable. Second, the Lagrangian of problem (5.25), i.e., $\mathcal{L}$ in (10.5), has a saddle point, since the Hessian of the Lagrangian is indefinite at the stationary point. The second condition is satisfied. As a result, the use of ADMM to solve problem (5.26) is convergent.

Complexity Analysis. For every iteration of ADMM, the complexity of evaluating the primal variable $\mathbf{\Lambda}_1$ is dominated by matrix multiplications and eigenvalue decomposition with a cost of $\mathcal{O}(K^3)$. Likewise, the cost of evaluating the primal variable $\mathbf{\Lambda}_2$ is $\mathcal{O}((N - K - 1)^3)$. The complexity of evaluating the primal variable $\mathbf{C}$ is dominated by matrix multiplications with a cost of $\mathcal{O}(N^2 K + N^2(N - K - 1))$. The update of the dual variable $\mathbf{Z}$ incurs the complexity of $\mathcal{O}(N^2 K + N^2(N - K - 1))$. Due to $K < N$, the overall cost is $\mathcal{O}(N^3)$ per iteration.

Remark 4 *The proposed approach can be extended to the case with observation noises. In this case, the observation signal, denoted by $\hat{\mathbf{Y}}$, is $\hat{\mathbf{Y}} = \mathbf{Y} + \mathbf{n}$, where $\mathbf{n} \sim \mathcal{N}(0, \sigma^2 \mathbf{I})$ is the additive white Gaussian noise and σ^2 is the variance of the additive white Gaussian noise. When the underlying signals $\mathbf{Y}$ are ergodic and M is sufficiently large, the expectation of $\hat{\mathbf{Y}} \hat{\mathbf{Y}}^T$ over the additive Gaussian observation noises is given by*

$$\mathbb{E}(\hat{\mathbf{Y}} \hat{\mathbf{Y}}^T) = \mathbf{Y}\mathbf{Y}^T + M \sigma^2 \mathbf{I}, \tag{5.27}$$

which indicates that $\mathbb{E}(\hat{\mathbf{Y}} \hat{\mathbf{Y}}^T)$ and $\mathbf{Y}\mathbf{Y}^T$ have the same eigenvectors.

Without observation noises, the GFT basis $\mathbf{U}^*$ in (5.14) only depends on the eigenvectors of $\mathbf{Y}\mathbf{Y}^T$. With the observation noises, the GFT basis is

$$\mathbf{U}^* = \text{eigen}\left[\left(\mathbf{I} - \mathbf{u}_1 \mathbf{u}_1^T \right) \mathbb{E}(\hat{\mathbf{Y}} \hat{\mathbf{Y}}^T) \left(\mathbf{I} - \mathbf{u}_1 \mathbf{u}_1^T \right)^T \right]. \tag{5.28}$$

Since $\mathbb{E}(\frac{1}{M} \hat{\mathbf{Y}} \hat{\mathbf{Y}}^T - \sigma^2 \mathbf{I}) = \frac{1}{M} \mathbf{Y}\mathbf{Y}^T$ (i.e., $\frac{1}{M} \hat{\mathbf{Y}} \hat{\mathbf{Y}}^T - \sigma^2 \mathbf{I}$ is an asymptotic unbiased estimate of $\frac{1}{M} \mathbf{Y}\mathbf{Y}^T$ as $M \to \infty$) based on (5.27) and $\hat{\mathbf{Y}} \hat{\mathbf{Y}}^T - M \sigma^2 \mathbf{I}$ and $\hat{\mathbf{Y}} \hat{\mathbf{Y}}^T$ have the same eigenvectors, $\mathbf{U}^*$ can be approximated by

$$\mathbf{U}^* \approx \text{eigen}\left[\left(\mathbf{I} - \mathbf{u}_1 \mathbf{u}_1^T \right) \hat{\mathbf{Y}} \hat{\mathbf{Y}}^T \left(\mathbf{I} - \mathbf{u}_1 \mathbf{u}_1^T \right)^T \right]. \tag{5.29}$$

Regarding the eigenvalues of $\mathbf{L}$, *i.e.*, $\boldsymbol{\Lambda}$, *the estimation with noise-free observations is formulated in* (5.17). *With noisy observation* $\hat{\mathbf{Y}}$, *an asymptotic unbiased estimate of the covariance matrix of* $\mathbf{Y}$, *i.e.*, $\lim\limits_{M\to\infty}\frac{1}{M}\mathbf{Y}\mathbf{Y}^T$, *is* $\frac{1}{M}\hat{\mathbf{Y}}\hat{\mathbf{Y}}^T - \sigma^2\mathbf{I}$, *since* $\lim\limits_{M\to\infty}\mathbb{E}\left(\frac{1}{M}\hat{\mathbf{Y}}\hat{\mathbf{Y}}^T - \sigma^2\mathbf{I}\right) = \lim\limits_{M\to\infty}\mathbb{E}\left(\frac{1}{M}\mathbf{Y}\mathbf{Y}^T\right)$ *[126]*. $\mathbf{Y}\mathbf{Y}^T$ *in* (5.17) *can be replaced with* $\hat{\mathbf{Y}}\hat{\mathbf{Y}}^T - M\sigma^2\mathbf{I}$. *In turn,* $\mathbf{T}$ *in* (5.19) *is updated to* $\hat{\mathbf{T}} = \hat{\mathbf{Y}}\hat{\mathbf{Y}}^T + (2\alpha - M\sigma^2)\mathbf{I}$ *under noisy observations and the rest of the estimation steps developed in this section apply.*

It is worth mentioning that **Remark 4** is based on the assumption that the observation noise follows a white Gaussian distribution. If the noise is colored with an unknown and non-diagonal covariance matrix, according to matrix perturbation theory [103], small perturbations on (the elements of) a matrix may result in considerable changes in its eigenvalues and eigenvectors. In this sense, the proposed algorithm could be sensitive to non-Gaussian observation noises.

5.6 Simulations and Experimental Results

Extensive experiments are performed to evaluate the new graph learning technique. The ADMM algorithm is initialized by setting $\mathbf{C}^{(0)}$ and $\mathbf{Z}^{(0)}$ to two symmetric unit matrices, and the step size ρ to 1. The algorithm is stopped upon reaching a predefined maximum number of iterations, i.e., 10^4, or the difference of the objective (5.26) is smaller than $\xi = 10^{-5}$ between two consecutive iterations. For comparison purposes, the following state of the arts are considered: Dong's algorithm [44], Saboksayr's algorithm [135], Sardellitti's Total Variation (TV) and Estimated-Signal-Aid (ESA) algorithms [140], and Egilmez's algorithm [52, 51].

5.6.1 Results on Synthetic Data

Three different random graphs generated separately from Random Geometric model [37], Erdős-Rènyi model [55], and Barábasi-Alber model [11] are considered. Regarding the Random Geometric model, the graph with six connections per node is generated. For the Erdős-Rènyi model, the graph by randomly connecting labeled nodes is generated. Each edge is included in the graph independently with a probability of 0.3, regardless of any other edge. For the Barábasi-Albert model, the graph is generated by starting with two initial nodes and then adding more nodes, each with two edges, as new nodes are added. For each of the three graph models, over 100 independent random graphs are generated. The average results of these graphs are then computed and plotted in this section.

After generating a graph from the graph models, the ground-truth Laplacian of the graph is obtained, denoted by $\mathbf{L}_0$. The singular value decomposition of $\mathbf{L}_0$ is taken to obtain the ground-truth GFT, denoted by $\mathbf{U}_0$. Then, the observed band-limited graph signal $\mathbf{Y} = \mathbf{U}_0\mathbf{S}_0$ with $\mathbf{S}_0 = [\mathbf{s}_{0,1}, \cdots, \mathbf{s}_{0,M}] \in \mathbb{R}^{N \times M}$ is generated. Specifically, $\mathbf{s}_{0,m} \sim \mathcal{N}(0, \mathbf{\Lambda}^\dagger)$, where $\mathrm{diag}(\mathbf{\Lambda}) = (\lambda_1, \cdots, \lambda_K, 0, \cdots, 0)$. The precision matrix of $\mathbf{s}_{0,m}$ is defined to be the eigenvalue matrix of $\mathbf{L}$ with the largest $(N - K)$ values set to 0, as considered in [91]. The key difference between the band-limited signals and smooth signals is the degenerate values in some dimensions of the multivariate Gaussian signal $\mathbf{S}$. The smooth signals can be considered as a specific scenario of the band-limited signals with $K = N - 1$. The presented algorithm, designed for general band-limited signals, can be applied to smooth signals.

With reference to [140], the performance metrics are F-measure, Recall, Precision score, and the percentage of Recovery errors. Let $\mathcal{E}_g$ and $\mathcal{E}_r$ denote the sets of ground truth and recovered graphs, respectively. Precision measures the proportion of the identified edges in the recovered graphs among the ground-truth graphs, i.e., Precision $= \mathcal{E}_g \cap \mathcal{E}_r / \mathcal{E}_r$. Recall evaluates the proportion of edges from the ground-truth graphs that are correctly identified among the recovered graphs, i.e., Recall $= \mathcal{E}_g \cap \mathcal{E}_r / \mathcal{E}_g$. F-measure combines both Precision and Recall into a single metric to evaluate the overall accuracy of the recovered edges. Specifically, F-measure $= 2 \cdot \text{Precision} \cdot \text{Recall} / (\text{Precision} + \text{Recall})$. The correlation coefficient $\rho_{\mathbf{W}}(\mathbf{W}_0, \mathbf{W})$ (or $\rho_{\mathbf{W}}$) between a recovered graph and its ground-truth is defined as $\rho_{\mathbf{W}}(\mathbf{W}_0, \mathbf{W}) = \frac{\sum_{ij} W_{0ij} W_{ij}}{\sqrt{\sum_{ij} W_{0ij}^2} \sqrt{\sum_{ij} W_{ij}^2}}$ [140], where $\mathbf{W}_0$ and $\mathbf{W}$ are the weighted adjacency matrices of the ground-truth and recovered graphs, respectively; and W_{0ij} and W_{ij} are the (i, j)-th elements of $\mathbf{W}_0$ and $\mathbf{W}$, respectively. The estimation error, or "Error," is defined as Error $= \|\mathbf{A} - \mathbf{A}_0\|_F / (N(N - 1))$, where $\mathbf{A}$ and $\mathbf{A}_0$ are the binary adjacency matrices of the recovered and ground-truth graphs, respectively.

Fig. 5.1(a) illustrates the convergence behaviors of the proposed algorithm under three different models. Fig. 5.1(b) plots the correlation coefficient of the algorithm, i.e., $\rho_{\mathbf{W}}$, with different values of the regularization parameter α and signal bandwidth K under the Random Geometric model. It can be seen that $\rho_{\mathbf{W}}$ reaches its peak at $\alpha = 0.9$ and $K = 15$; i.e., the optimal regularization parameter is $\alpha = 0.9$ when the signal bandwidth is $K = 15$. Likewise, the optimal α under the Erdős-Rènyi and Barábasi-Albert models is obtained.

Tables 5.1 and 5.2 compare the proposed algorithm and the benchmarks under the considered graph models, where each random graph consists of $N = 30$ vertices. $M = 300$. K is set to 3 in Table 5.1. In Table 5.2, K is set to 15. Different values of the regularized parameter α are tried and tested to obtain the best one, as done in Fig. 5.1(b). For fair comparisons, α is also individually tested and optimized for each benchmark. The tables show that the proposed algorithm consistently outperforms all the other techniques considered, indicating

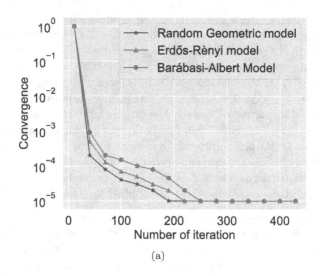

(a)

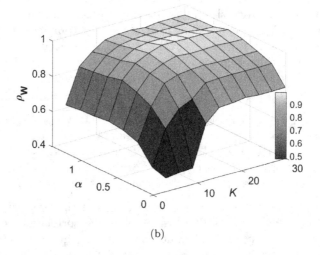

(b)

FIGURE 5.1: (a) The convergence of the proposed algorithm. (b) The correlation coefficient vs. K and α under the Random Geometric model, where $N = 30$ and $M = 300$.

its superiority in terms of performance. It achieves high scores in both F-measure and correlation coefficient $\rho_{\mathbf{W}}$ under all random graph models. In particular, Table 5.1 shows that the proposed algorithm yields at least 20% improvement in terms of F-measure under the Erdős-Rènyi model, compared to Sardellitti's TV and ESA algorithms. By comparing Tables 5.1 and 5.2, It shows that the proposed algorithm increasingly outperforms the benchmarks, as K grows.

Table 5.1: Comparison of the considered algorithms, including the new algorithm, Dong's [44], Saboksayr's (Sab.) [135], Sar-TV [140], Sar-ESA [140], and Egilmez's [52, 51], where $K = 3$, $N = 30$, and $M = 300$.

	Dong	Sab.	TV	ESA	Egilmez	New Algorithm
RG						
F-measure	0.5478	0.5499	0.5536	0.5255	0.4770	**0.5597**
Recall	0.4500	0.4476	**0.4627**	0.3913	0.3217	0.4213
Precision	0.7000	0.7129	0.6889	0.8000	**0.9222**	0.8333
ρ_W	0.6324	0.5357	0.5996	0.5600	0.6115	**0.6471**
Error	0.0189	0.0185	0.0220	0.0219	0.0261	**0.0177**
ER						
F-measure	0.4632	0.4971	0.3784	0.3033	0.2876	**0.5283**
Recall	0.5000	**0.5096**	0.2612	0.2200	0.1725	0.3889
Precision	0.4314	0.4852	0.6863	0.6275	0.8627	**0.8235**
ρ_W	0.5611	0.5761	0.4623	0.4496	0.5670	**0.6215**
Error	**0.0170**	0.0199	0.0174	0.0197	0.0200	0.0189
BA						
F-measure	**0.5056**	0.5014	0.3784	0.3474	0.3803	0.4808
Recall	**0.3980**	0.3999	0.2612	0.2357	0.2368	0.3289
Precision	0.6946	0.6717	0.6863	0.6607	0.9643	**0.8929**
ρ_W	0.4495	0.4486	0.4221	0.3054	0.4151	**0.5135**
Error	**0.0142**	0.0173	0.0177	0.0192	0.0216	0.0169

Table 5.2: Comparison of the considered algorithms, where $K = 15$, $N = 30$, and $M = 300$.

	Dong	Sab.	TV	ESA	Egilmez	New Algorithm
RG						
F-measure	0.8517	0.9111	0.7946	0.6642	0.9301	**0.9401**
Recall	0.7479	0.8427	0.6666	0.4972	0.8693	**0.8912**
Precision	0.9889	0.9914	0.9889	0.9999	1.0000	**1.0000**
ρ_W	0.9156	0.9342	0.9299	0.9442	0.9641	**0.9895**
Error	0.0091	0.0102	0.0190	0.0179	0.0136	**0.0075**
ER						
F-measure	0.8523	0.8671	0.5556	0.6221	0.8963	**0.9254**
Recall	0.8009	0.8183	0.4005	0.4728	0.8425	**0.8907**
Precision	0.9107	0.9221	0.9020	0.9092	0.9575	**0.9610**
ρ_W	0.8984	0.9207	0.9289	0.9221	0.9244	**0.9840**
Error	0.0064	0.0059	0.0145	0.0185	0.0115	**0.0083**
BA						
F-measure	0.7257	0.7857	0.5368	0.4250	0.4576	**0.8728**
Recall	0.6164	0.6986	0.3806	0.2772	0.3000	**0.8359**
Precision	0.8836	0.8976	0.9107	0.9109	0.9643	**0.9131**
ρ_W	0.7198	0.7371	0.8076	0.8074	0.7962	**0.9641**
Error	0.0117	0.0153	0.0162	0.0191	0.0184	**0.0105**

Fig. 5.2(a) evaluates the average correlation coefficients of the considered algorithms against K under the Random Geometric model, where $N = 30$ and $M = 300$. The presented algorithm achieves the largest correlation coefficient over the wide spectrum of K. It can be seen that the correlation coefficient first increases with K, reaches its peak at $K = 15$, and then stabilizes. This is

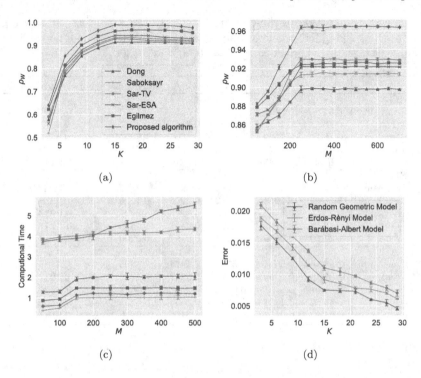

(a) (b)

(c) (d)

FIGURE 5.2: (a) The correlation coefficient with the increased signal bandwidth K under the Random Geometric model, where $N = 30$ and $M = 300$. (b) The correlation coefficient with the increase of M under the Erdős-Rènyi model, where $K = 15$ and $N = 30$. (c) Runtime (in seconds) with the increase of M under the Random Geometric model, where $K = 15$ and $N = 30$. (d) The estimation error of the proposed algorithm with the increased signal bandwidth K, where $N = 30$ and $M = 300$.

because a wider signal bandwidth K allows for a broader range of frequencies to be captured in the observed signals. As a result, more detailed and diverse information can be extracted. When K exceeds 15, the information extracted from the observed signals no longer increases because the observed signals are sparse, and the setting of $K \geq 15$ overkills. Fig. 5.2(b) compares the correlation coefficients between different methods under the Erdős-Rènyi model, as M increases. $K = 15$. It shows that the correlation coefficients stabilize when $M > 250$. Fig. 5.2(c) compares the computational complexities of the methods with the increase of M, where the Random Geometric model is considered, $K = 15$, and the average of 1,000 independent tests are plotted. It shows that Saboksayr's method is the fastest, followed by the proposed algorithm. This indicates that the algorithm provides superb detection accuracy with considerably low complexity. Fig. 5.2(d) illustrates that the Error score consistently decreases as the value of K increases across all three models.

5.6.2 Experiment on Temperature Measurement Data

The performance of the proposed graph learning algorithm using a temperature dataset collected from 32 weather stations in Brittany, France[1] is assessed. The dataset collects hourly temperatures during a period of 31 days, resulting in a total of $24 \times 31 = 744$ observations. The 32 weather stations are taken as the vertices of a graph and the temperatures measured at the stations as the observation $\mathbf{Y} \in \mathbb{R}^{32 \times 744}$. Given $\mathbf{Y}$, the GFT basis is first analytically determined, $\mathbf{U}$, using (5.14) in Section 5.3 and then recover the graph using the ADMM algorithm developed in Section 5.4. In the recovered graph, the vertices indicate the weather stations. The weight of an edge can then indicate the degree of similarity between the temperatures of two weather stations represented by the connected vertices.

Fig. 5.3 shows the graphs learned, where the average temperature of each weather station is color-coded. Fig. 5.3(a) corresponds to the proposed algorithm, where $K = 17$ is obtained as described in **Remark 3**, i.e., by first setting $K = N - 1$ to generate and sort $\|\mathbf{S}(i, :)\|$, i.e., $\|\mathbf{S}(1, :)\| \geq \|\mathbf{S}(2, :)\| \geq \cdots \geq \|\mathbf{S}(N, :)\|$, and then selecting the smallest K such that $\sum_{i=1}^{K} \|\mathbf{S}(i, :)\| / \sum_{i=1}^{N} \|\mathbf{S}(i, :)\| \geq 99\%$. The results of the existing algorithms are plotted in Figs. 5.3(b)–5.3(d). For Sar-ESA and Sar-TV, $K = 17$ is obtained in the same way as in the presented algorithm. Saboksayr's and Egilmez's do not rely on the value of K. For fair comparisons, all graphs are configured to have consistent sparsity, i.e., each graph has 90 edges, by fine-tuning the respective sparsity hyperparameters of the considered algorithms. It can be seen that the proposed algorithm provides more reasonable learning results, as it connects the weather stations with close average temperatures and disconnects those with substantially different temperatures. By contrast, the existing approaches connect the weather stations excessively with substantially different average temperatures in Figs. 5.3(b), 5.3(d), and 5.3(e); or fail to connect some nearby weather stations with similar average temperatures in Fig. 5.3(c).

The reliability of the proposed algorithm is evaluated by using a greedy algorithm [23] to reconstruct the temperature data of some of the stations over a random period of time based on the learned Laplacian matrix and the observed signals of the other stations. The number of observable temperature signals (or weather stations) is set to be equal to the signal bandwidth K. In other words, 17 nodes are selected to recover the signal waveforms of the other 15 nodes. Fig. 5.4(a) shows the ground-truth temperatures observed at Lorient station (the solid blue line) and the reconstructed version (the orange dash line). The Lorient station is one of the 15 unobserved stations. It can be seen that the reconstructed signals are consistent with the ground-truth signals, indicating the reliability of the new graph-learning algorithm. For comparison, the temperature signals by running Saboksayr's [135], Sar-ESA [140], Sar-TV

---

[1]https://github.com/BaolingShan/Temperature-datasset-in-Brittany-France

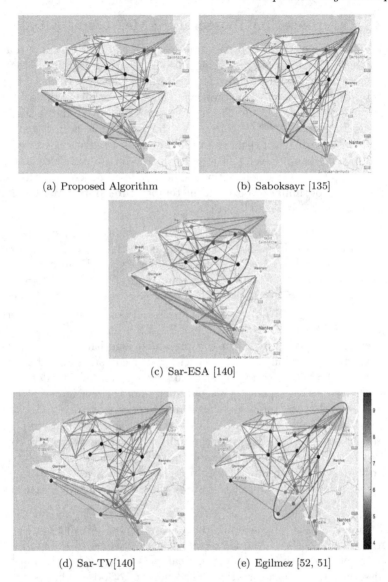

(a) Proposed Algorithm (b) Saboksayr [135]

(c) Sar-ESA [140]

(d) Sar-TV[140] (e) Egilmez [52, 51]

FIGURE 5.3: The learned graphs with different methods. The color of a node indicates the average temperature in $°C$ measured at the node during all the periods of observation.

[140], and Egilmez's [52, 51] are also reconstructed, as shown in Figs. 5.4(b)-5.4(e), where the reconstructed signals are less accurate. The coefficient of determination is adopted, denoted by R^2, to quantify the reconstruction accuracy of the algorithms. $R^2 = 1 - \|\text{vec}(\hat{\mathbf{Y}}_R) - \text{vec}(\mathbf{Y}_R)\|_2^2 / \|\text{vec}(\hat{\mathbf{Y}}_R) - \bar{y}_R\|_2^2$.

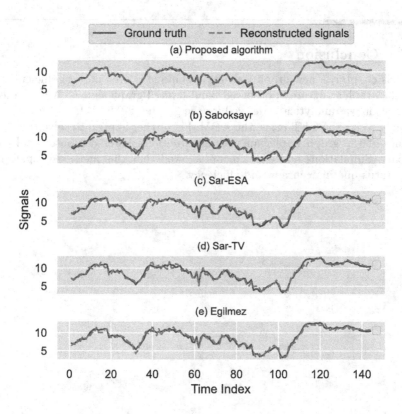

FIGURE 5.4: The reconstructed signals of temperatures at Louargat.

$\hat{\mathbf{Y}}_R$ and $\mathbf{Y}_R$ are the reconstructed signals and their ground-truths. $\bar{y}_R$ is the average of the ground-truths. $\text{vec}(\cdot)$ denotes vectorization. Table 5.3 shows this algorithm provides the best reconstruction accuracy with the second shortest runtime.

Table 5.3: The reconstruction accuracy of Saboksayr's (Sab.) [135], Sar-ESA [140], Sar-TV [140], Egilmez's [52, 51], and the proposed algorithm upon temperature dataset.

Methods	Sab.	Sar-ESA	Sar-TV	Egilmez	Proposed Algorithm
Accuracy	0.7227	0.8269	0.7343	0.7061	**0.9301**
Runtime (s)	15	203	156	37	23

5.7 Conclusion

In this chapter, a new graph learning technique is presented to efficiently infer the graph structure of observed band-limited graph signals. A closed-form derivation was analytically derived for the graph GFT basis from the observed signals. With the GFT basis, the ADMM is further applied to solve the eigenvalues of the graph Laplacian with substantially lower computational complexity. Simulations and experiments showed that this method outperforms the status quo in accuracy and efficiency.

6

Graph Topology Learning of Brain Signals

Graph theory has played a crucial role as a tool for analyzing intricate brain networks, and revealed several non-trivial features of brains, such as modularity and small-worldness, by studying the fMRI time series (i.e., blood-oxygen-level dependent time series) amongst the ROIs in a brain [115]. This chapter presents the application of the graph learning technique in Chapter 4, which learns weighted and undirected graph topologies, more specifically, the graph Laplacian matrices, from fMRI-based, band-limited brain signals.

6.1 Introduction

Network neuroscience contributes to the comprehension of the architecture and functionality of the human brain by viewing a brain to be a complex network comprising many ROIs, also known as brain network nodes [115]. The connectivity and functional interactions among the ROIs serve a key role in brain-related cognitive function [89]. One example is ADHD, a widely observed and severely impairing neurodevelopmental behavioral problem diagnosed with inattention, hyperactivity, and impulsivity. ADHD affects school-age children and adolescents [56]. Latest neuroanatomical and neuropsychological studies indicate that these behavioral disturbances relate to atypical connectivity amongst brain functional area [47].

Pearson's correlation has been one of the most common methods for measuring the pairwise functional relationships between brain regions. However, Pearson's correlation focuses on strong direct marginal correlations in the fMRI data among two brain regions and overlooks the latent effects of other brain regions. Albert *et al.* [4] used partial correlations to measure interactions between any two ROIs. The partial correlation quantifies the dependency between two ROIs by regressing out the other ROIs, leading to difficulties in the suppression of the confounding effect from the other ROIs. SICE [178] has been another popular technique for measuring the intensity of the most significant direct connection between ROIs. SICE is a rigorous algorithm based on partial correlation. It tends to evaluate the sparsest reconstructive coefficient

DOI: 10.1201/9781003516613-6

of each ROI and capture only local structures, rather than a representation of the global structure.

6.2   Materials and System Model

A whole brain network can be parcellated into a set of N nodes corresponding to different brain regions. Each node represents a specific region containing a fMRI time series. It assumes that the i-th time series corresponding to N nodes refers to a vector, denoted by $\mathbf{y}_i \in \mathbb{R}^{N \times 1}$. M is the number of sampling times of brain signals. Let $\mathbf{Y} = [\mathbf{y}_1, \ldots, \mathbf{y}_M] \in \mathbb{R}^{N \times M}$ collect M observed signals of the brain network.

Brain networks describe the physical connectivity patterns between different brain regions. As considered in [140], the objective is to deduce the graph topology of brains from the time series $\mathbf{Y}$. Specifically, it wishes to characterize the brain network with a weighted, undirected, and unidentified graph $\mathcal{G} = (\mathcal{V}, \mathcal{E}, \mathbf{W})$ consisting of N vertices. Every vertex corresponds to a brain network node. The edge between any two vertices indicates the physical proximity or relationship of the corresponding brain nodes. $\mathcal{V} = \{1, \cdots, N\}$ collects N vertices corresponding to specific brain regions. $\mathcal{E} \subseteq \mathcal{V} \times \mathcal{V}$ is the collection of edges of the brain connectivity. The adjacent matrix $\mathbf{W} \in \mathbb{R}^{N \times N}$ indicates to what extent two brain nodes are correlated. $W_{ij} = W_{ji} \neq 0$ for any brain regions $(i, j) \in \mathcal{E}$.

The combinatorial graph Laplacian $\mathbf{L}$ of the brain network $\mathcal{G}$ is defined as [146]:

$$\mathbf{L} = \mathbf{D} - \mathbf{W}, \tag{6.1}$$

where $\mathbf{D} \triangleq \mathrm{diag}\,(\mathbf{W1})$ defines the degree matrix containing the node degrees at its diagonal. $\mathbf{1}$ is an all-one vector. This chapter assumes that each brain region is connected to at least one other brain region, ensuring no isolated regions are in the brain topology. In other words, none of the diagonal elements is zero in $\mathbf{D}$.

The graph Laplacian is a semi-definite matrix with positive elements along its main diagonal and non-positive elements anywhere else [52]. By eigenvalue decomposition, $\mathbf{L}$ is rewritten as:

$$\mathbf{L} = \mathbf{U\Lambda U}^T, \tag{6.2}$$

where $\mathbf{\Lambda}$ denotes the diagonal matrix of non-negative Laplacian eigenvalues, and $\mathbf{U} = [\mathbf{u}_1, \cdots, \mathbf{u}_N]$ represents the orthonormal matrix collecting all the eigenvectors.

To infer the topological knowledge of $\mathcal{G}$ requires estimating the Laplacian matrix $\mathbf{L}$. As done in [140] and [91], this chapter enforces the signals $\mathbf{Y}$ to be band-limited over graph $\mathcal{G}$, e.g., the signals are sparse in the canonical domain [140, 91, 170]. GFT [146, 131] has been utilized to decompose a brain time series into orthonormal components $\mathbf{U}$ in the Laplacian $\mathbf{L}$ [52]. For any $m \in \{1, \cdots, M\}$, the GFT of the time series $\mathbf{y}_m$, denoted by $\mathbf{s}_m$, projects $\mathbf{y}_m$ onto spectral-domain subspace spanned by $\mathbf{U}$, as given by

$$\mathbf{s}_m = \mathbf{U}^T \mathbf{y}_m. \tag{6.3}$$

With the band-limitedness of the observed signal $\mathbf{y}_m$, $\mathbf{s}_m$ is a sparse vector and captures the key characteristics of $\mathbf{y}_m$ in the frequency domain. The band-limited signal is written as

$$\mathbf{y}_m = \mathbf{U}\mathbf{s}_m. \tag{6.4}$$

Let $\mathbf{S} = [\mathbf{s}_1, \cdots, \mathbf{s}_M] \in \mathbb{R}^{N \times M}$ collect all $\mathbf{s}_m \in \mathbb{R}^{N \times 1}$, $m \in \{1, \cdots, M\}$. From (6.4), it has

$$\mathbf{Y} = \mathbf{U}\mathbf{S}. \tag{6.5}$$

With the sparsity of $\mathbf{s}_m, m \in \{1, \cdots, M\}$, it sets $\mathbf{S} \in \mathcal{B}_K$ as a K-block sparse matrix with multiple all-zero row-vectors. K accounts for the bandwidth of the frequency-domain representation of the observed band-limited graph signals $\mathbf{Y}$, which can be obtained empirically in prior, or enumerated to find its proper value. $\mathcal{B}_K$ collects all K-block sparse matrices [53]:

$$\mathcal{B}_K \triangleq \{\mathbf{S} \in \mathbb{R}^{N \times M}, \mathbf{S}(i, :) = \mathbf{0}, \forall i \notin \mathcal{K} \subseteq \mathcal{V}, K = |\mathcal{K}|\}, \tag{6.6}$$

where $\mathbf{S}(i, :)$ is the i-th row of $\mathbf{S}$, and $\mathcal{K} \subseteq \mathcal{V}$ has the cardinality of K.

6.3 Graph Inference for ADHD Data

This chapter applies Algorithm 1 to analyze brain functional connectivity networks and demonstrate the effectiveness of the algorithm compared to existing techniques with respect to reliability and efficiency. The considered ADHD dataset contains 42 right-handed male subjects aged between 11 and 16 years old[1]. To analyze the connectivity of the brain functional networks, this scheme divides a brain into 90 anatomical ROIs using an anatomical automatic labeling template [155], where each node accounts for an ROI with 232 signals from the ADHD dataset. The observed brain signals yield $\mathbf{Y} \in \mathbb{R}^{90 \times 232}$.

---

[1]The dataset is obtained from the ADHD-200 global competition database (https://www.nitrc.org/projects/neurobureau/). The dataset contains 17 ADHD subjects and 25 TD subjects that are analyzed and compared in this section.

Since the actual graphs underlying the ADHD data are unknown, an alternate way must be found to evaluate the accuracy of Algorithm 1 in extracting the graphs of the ADHD dataset. In this case, it relies on evaluating the learning accuracy of the algorithm. To do this, it first sets $K = N$ to calculate and arrange $\|\mathbf{S}(i,:)\|$ in descending order, i.e., $\|\mathbf{S}(1,:)\| \geq \|\mathbf{S}(2,:)\| \geq \cdots \geq \|\mathbf{S}(N,:)\|$, and then selecting the minimum K that $\frac{\sum_{i=1}^{K} \|\mathbf{S}(i,:)\|}{\sum_{i=1}^{N} \|\mathbf{S}(i,:)\|} \geq 99\%$. The resultant $K = 62$ is used to learn the graph of the ADHD data.

The brain topology is displayed by the BrainNet Viewer toolbox [168]. This chapter considers that the ROIs belong to eight anatomical regions, i.e., frontal lobe (dark blue), orbital surface (blue-green), temporal lobe (green), parietal lobe (blue), occipital lobe (orange), limbic lobe (yellow), insula (light green), and subcortical gray nuclei (red); see Fig. 6.1.

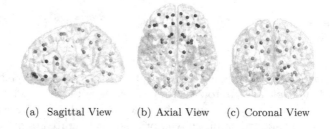

(a) Sagittal View   (b) Axial View   (c) Coronal View

FIGURE 6.1: Colored nodes distribution from sagittal, axial, and coronal views. Different colors of nodes represent different blocks of ROIs.

Fig. 6.2 shows the variations in the brain functional networks between ADHD subject 17 (ADHD-17) and typical developing (TD) subject 23 (TD-23), which are visualized from the sagittal, axial, and coronal views of the brain. It is observed that both ADHD and TD subjects exhibit efficient small-world brain network structures and have highly similar hub distributions. Nevertheless, ADHD-17 exhibits substantial asymmetry in brain anatomical network topology, as shown in the sagittal, axial, and coronal views in Figs. 6.2. It also can be observed that ADHD-17 shows reduced structural connectivity and weakened connection strength within the same anatomical regions, especially in the frontal lobe (dark blue nodes), temporal lobe (green nodes), parietal lobe (blue nodes), and occipital lobe (orange nodes), compared to TD-23. Moreover, ADHD-17 has fewer connections between different anatomical regions than TD-23, which is consistent with existing neuroanatomical studies [50].

To better display the difference in the brain functional networks between the ADHD and TD subjects, it transforms the brain functional network into the weighted adjacency matrices of ADHD-17 and TD-23 in Fig. 6.3. It is found that ADHD-17 exhibits a considerably varying degree of decreased connectivity, compared to TD-23. The observed alterations in ADHD-17 provide further evidence that the brain network structures of individuals with ADHD change

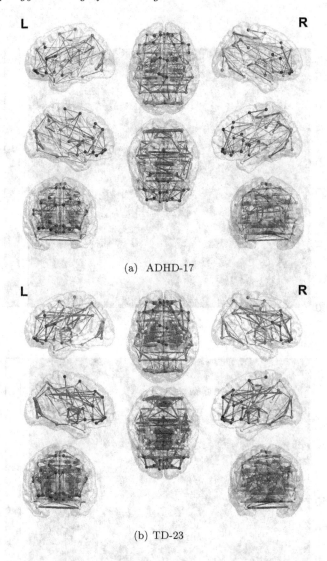

(a) ADHD-17

(b) TD-23

FIGURE 6.2: Illustration of connected brain networks from different perspectives using BrainNet Viewer. Colored nodes denote different blocks of ROIs, and the intensity of the red lines signifies the connectivity strength between two ROIs. (a) The learned graphs from ADHD-17 using Algorithm 1. (b) The learned graph from TD-23.

in distributed neural networks, potentially contributing to symptoms of inattention and hyperactivity. The observable decreased structural connectivity within the local functional networks of ADHD-17, and increased asymmetry globally in the brain functional network of ADHD-17 align with findings from previous neuroanatomical studies, e.g., [71, 50, 19, 93, 49].

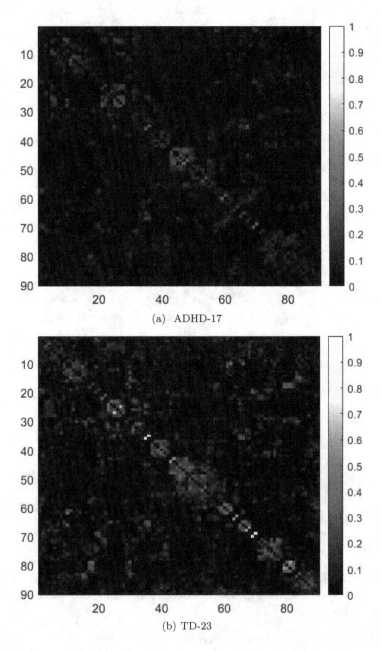

(a) ADHD-17

(b) TD-23

FIGURE 6.3: The weighted adjacency matrix of the learned graph based on the ADHD-17 and TD-23.

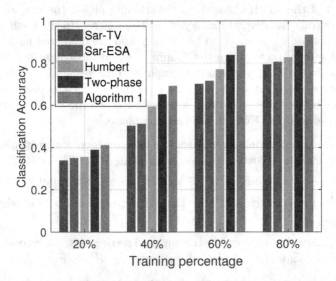

FIGURE 6.4: The classification accuracy of different methods under different ratios between the training and testing sets.

This chapter also conducts classification based on the visualized weighted adjacency matrices of all subjects in the ADHD dataset (including 17 ADHD subjects and 25 TD subjects). A Support Vector Machine (SVM) classifier is employed [117]. As shown in Fig. 6.4, the classification accuracy is compared between the weighted adjacency matrices generated by the proposed algorithm, and the benchmarks, i.e., Sar-TV [140], Sar-ESA [140], Humbert [79], and Two-phase algorithm [143], across various ratios between the training and testing sets. When the training set is composed of 80% of the data, and the testing set comprises 20% of the data, Algorithm 1 achieves the highest classification accuracy with improvements of approximately 4.78%, 12.91%, 15.83%, and 17.65%, compared to Two-phase algorithm [143], Humbert [79], Sar-ESA [140], and Sar-TV [140], demonstrating the practical value of the proposed algorithm for the diagnosis of brain disorders, e.g., ADHD.

6.4   Method Assessment and Results

To further gauge the accuracy of Algorithm 1 in Chapter 4, experiments are carried out to test the capabilities of the considered algorithms in learning graph structures from a training set and then predicting missing data/observations in a test set. The ADHD-17 set is divided in the time domain between a training set (i.e., the early part of the ADHD-17 set) and a testing set (i.e., the late part of the ADHD-17 set). In the training phase, the graph learning algorithms extract the graph topology of the training set. In the testing phase, the data of node 9 are assumed to be missing, and are predicted based on the graph topologies extracted from the training set and the available data of the other nodes. In addition to the latest graph learning techniques, this section takes the state-of-the-art GCN [181] as a benchmark for the proposed algorithm, where it constructs the brain network based on the pairwise Pearson's correlation between the ROIs in the training phase.

To measure the reconstruction efficiency of the considered algorithms, it utilizes the coefficient of determination, represented by R^2, i.e., $R^2 = 1 - \frac{\Sigma_{i=1}^{N}\Sigma_{j=1}^{M}(P_{ij}-Y_{ij})^2}{\Sigma_{i=1}^{N}\Sigma_{j=1}^{M}(P_{ij}-\bar{Y}_i)^2}$, where P_{ij} and Y_{ij} are the reconstructed and ground-truth brain signals, respectively; and $\bar{Y}_i$ is the mean ground-truth brain signals at the i-th node.

By adjusting the ratio between the training and testing sets, the reconstruction accuracy of the algorithms is plotted, measured by R^2, in Fig. 6.5.

As illustrated in Fig. 6.5, the graph learning methods, including Algorithm 1, demonstrate superior performance compared to the GCN across various ratios between the training and testing sets. Notably, when the proportion of the training set is set to 80%, and the testing set is 20%, Algorithm 1 attains the highest R^2 values, showcasing an improvement of 5.39%, 14.58%, 15.63%,

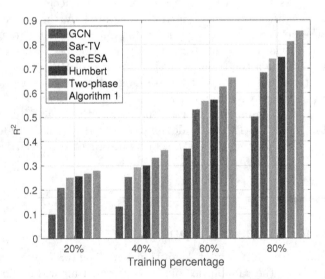

FIGURE 6.5: The reconstruction accuracy R^2 of the considered methods under different ratios between the training and testing sets.

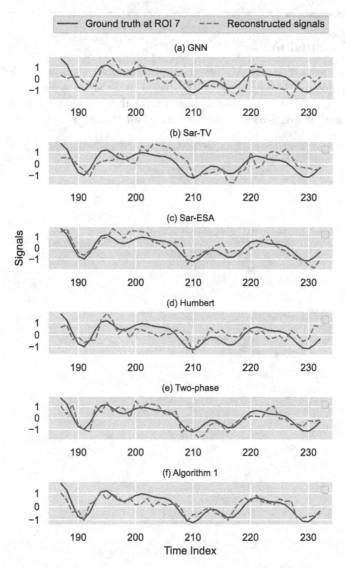

FIGURE 6.6: The signals reconstructed for the anatomical ROI 9 from ADHD-17 using the different methods, where the training set is 80% of the ADHD-17 data set and the testing set is the remaining 20% of the data set.

and 67.13%, compared to Two-phase [143], Humbert [79], Sar-ESA [140], Sar-TV [140], and GCN [181], respectively. This demonstrates the applicability of graph learning techniques to situations where training data is scarce, compared to the GCN. The reconstructed signals of different methods are shown in Fig. 6.6.

6.5 Conclusion

This paper presented the AO-based graph learning technique to deduce the graph topology of band-limited, fMRI-based brain signals. Experiments corroborated that Algorithm 1 in Chapter 4 can substantially improve the learning accuracy of fMRI-based brain signals than the state of the art, e.g., by over 30%.

7

Graph Topology Learning of COVID-19

This chapter presents a new graph-learning technique to accurately infer the graph structure of coronavirus disease 2019 (COVID-19) data, helping to reveal the correlation of pandemic dynamics among different countries and identify influential countries for pandemic response analysis. The new technique estimates the graph Laplacian of the COVID-19 data by first deriving analytically its precise eigenvectors, also known as GFT basis. Given the eigenvectors, the eigenvalues of the graph Laplacian are readily estimated using convex optimization. Using the graph Laplacian, the analysis of confirmed cases of different COVID-19 variants among European countries is conducted based on centrality measures. This approach identifies a different set of the most influential and representative countries compared to current techniques. The accuracy of the new method is validated by repurposing part of COVID-19 data to be the test data and gauging the capability of the method to recover missing test data, showing 33.3% better in root mean squared error (RMSE) and 11.11% better in correlation of determination than existing techniques. The set of identified influential countries by the method is anticipated to be meaningful and contribute to the study of COVID-19 spread.

7.1 Introduction

Global health, economic, and social challenges have been escalating since the COVID-19 pandemic. As of April 2022, Europe had 192.09 million confirmed cases and over 2 million deaths [1]. The SARS-CoV-2 virus has undergone numerous genetic changes since its discovery [165]. While some of these changes do not affect the virus's behavior, others may affect how easily it is transmitted. Changes beneficial to the virus tend to spread more quickly, which means that variants harboring them gradually replace other circulating variants [166]. In November 2020, SARS-CoV-2 Alpha was first identified in the United Kingdom, which was estimated to have 50% higher transmissibility than the original strain. From July 2021 to October 2021, SARS-CoV-2 Delta prevailed in Europe. The SARS-CoV-2 Omicron variant took over from the SARS-CoV-2

---

[1] https://coronavirus.jhu.edu/data

DOI: 10.1201/9781003516613-7

Delta variant in Europe in November 2021. Earlier studies demonstrated that Omicron can, to a degree, evade the protective effects of antibodies induced by vaccinations or natural infections. Large portions of the European population are susceptible to infection, leading to sharp increases in COVID-19 cases and unprecedented community spread.

Comprehending the spatio-temporal characteristics of the virus transmission is the key to controlling the transmission of the pandemic. Studies show that the global spread of COVID-19 did not process uniformly [66, 153]. An outbreak's size and condition are influenced by the characteristics of virus spread [68]. Unfortunately, it is difficult to implement evidence-based policies for COVID-19 due to a lack of adequate evidence in policy-making and research [57]. While it is possible to estimate the growth rates of confirmed cases and deaths [46], the relationships between pairs of countries are still unknown as far as COVID-19 development is concerned. Datasets about ongoing situations in different countries are likely to show spatial-temporal patterns since the spread of the virus tends to follow geographic trends. A spatial-temporal analysis of confirmed COVID-19 cases may also shed light on its evolution. The record of pandemic evolution in Europe is known to be complex, variable, and non-linear. Consequently, it is essential to uncover hidden information about SARS-CoV-2 as new virus variants emerge.

One way to understand the spreading dynamic of the pandemic is to generate and analyze COVID-19 pandemic diffusion graph topologies with the graph-theoretic metrics [90, 119, 39, 147]. In addition to illustrating spatial and temporal connections between places, spatio-temporal maps can potentially indicate changes in pandemic risks [130].

Existing studies have examined the spread of epidemics as a complex system by assessing the degree of correlation or synchronization between time-series data [88]. A deeper understanding of the spread dynamics of the new variants of SARS-CoV-2 requires new methods beyond assessing correlation or synchronization. There is a need to explore the latent structures among the data and reveal the relationships between different countries to understand the spatio-temporal spread of the virus.

7.2   Materials and System Model

The analysis is based on the open-access dataset of daily identified COVID-19 cases reported officially by different countries, territories, and regions, and published by the WHO[2]. The daily data for COVID-19 in European countries are updated every day. The data is collected from January 2020 to April 2022

---

[2]https://covid19.who.int/WHO-COVID-19-global-data.csv

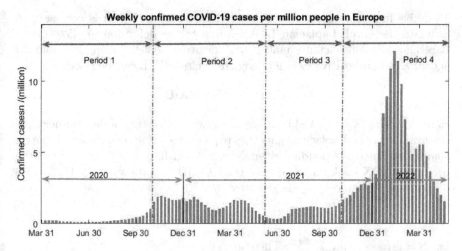

FIGURE 7.1: Weekly confirmed COVID-19 cases per million people.

and divided this period into four based on the statistics from the WHO, as shown in Fig. 7.1. The first period is the early stage of the pandemic outbreak, between March 2020 and October 2020, when the original strain of the virus dominated the spread. The second period is from November 2020 to May 2021, when the Alpha variant was dominant. The third stage is from June 2021 to October 2021, when the Delta variant broke out. The fourth stage is from November 2021 to April 2022, when the Omicron variant rapidly replaced the Delta and became the dominating variant in most European countries.

For each period, the SARS-CoV-2 time series data of the 44 European countries, published by the WHO, is analyzed by extracting a graph with 44 nodes from the data. The graph topology is defined as $\mathcal{G} = (\mathcal{V}, \mathcal{E}, \mathbf{W})$, where $\mathcal{V} = 1, 2, \ldots, N$ denotes the set of N vertices, with $N = 44$ being the number of countries. The set of edges, denoted by $\mathcal{E}$, is a subset of $\mathcal{V} \times \mathcal{V}$. The weighted adjacency matrix of the graph $\mathcal{G}$, denoted by $\mathbf{W} \in \mathbb{R}^{N \times N}$, indicates the extent to which two countries are correlated with respect to COVID-19 spread. $W_{ij} = W_{ji} \neq 0$ for $\forall (i, j) \in \mathcal{E}$. Each node in $\mathcal{G}$ represents a European country and corresponds to the time series recording daily confirmed cases per million people in the country.

For each of the periods, $\mathbf{x}_p \in \mathbb{R}^{N \times 1}, \forall p \in \{1, \cdots, P\}$ denotes the COVID-19 records of the N countries on the p-th day of the period, where P is the number of days in the period. The COVID-19 data of the European countries during the period are arranged in an $N \times P$ matrix, denoted by $\mathbf{X} = [\mathbf{x}_1, \ldots, \mathbf{x}_P] \in \mathbb{R}^{N \times P}$. Here, $\mathbf{X}$ is band-limited and its frequency-domain representation has finite bandwidth; in other words, the virus spreads across countries, rather than breaks out simultaneously in all countries.

To derive information about the underlying structure of $\mathcal{G}$, it is necessary to estimate the graph Laplacian $\mathbf{L}$. According to the definition in [52], graph Laplacian is a semi-definite matrix with positive elements along its main diagonal and non-positive elements anywhere else, which can be rewritten as:

$$\mathbf{L} = \mathbf{U}\boldsymbol{\Lambda}\mathbf{U}^T = \mathbf{U}\text{diag}(\boldsymbol{\lambda})\mathbf{U}^T. \tag{7.1}$$

Here, $\boldsymbol{\Lambda} = \text{diag}(\lambda_1, \cdots, \lambda_N)$ is a diagonal matrix consisting of the non-negative eigenvalues of the Laplacian, and $\mathbf{U} = [\mathbf{u}_1, \cdots, \mathbf{u}_N]$ is an orthonormal matrix comprising the corresponding eigenvectors. According to [23], the GFT is the projection of $\mathbf{X}$ on $\mathbf{U}$. The GFT of the COVID-19 data $\mathbf{x}_p, \forall p \in \{1, \cdots, P\}$ on p-th day, denoted by $\mathbf{s}_p$, is given by

$$\mathbf{s}_p = \mathbf{U}^T\mathbf{x}_p. \tag{7.2}$$

Let $\mathbf{S} = [\mathbf{s}_1, \cdots, \mathbf{s}_p] \in \mathbb{R}^{N \times P}$. Then,

$$\mathbf{X} = \mathbf{U}\mathbf{S}. \tag{7.3}$$

With the sparsity of $\mathbf{s}_p, \forall p$, $\mathbf{S}$ is defined as a K-block sparse matrix with rows consisting of multiple all-zero vectors. Here, K indicates the frequency-domain bandwidth of the COVID-19 data $\mathbf{X}$. K is obtained empirically in prior or enumerated to find its proper value [12]. The set $\mathcal{B}_K$ contains all K-block sparse matrices, defined as $\mathcal{B}_K \triangleq \{\mathbf{S} \in \mathbb{R}^{N \times P}, \mathbf{S}(i,:) = \mathbf{0}, \forall i \notin \mathcal{K} \subseteq \mathcal{V}, K = |\mathcal{K}|\}$. Here, $\mathbf{S}(i,:)$ denotes the i-th row of $\mathbf{S}$, and the set $\mathcal{K} \subseteq \mathcal{V}$ collates the indexes to the K most significant frequency components of the $\mathbf{X}$.

7.3 Proposed Graph Inference for COVID-19 Spread Analysis

COVID-19 data analysis plays a crucial role in identifying the most influential countries or regions in the spread of the virus and understanding how the virus spreads among countries. In this section, a new graph learning method is developed, which accurately and efficiently extracts the underlying graph topological information of the COVID-19 data. This method reveals the fine-grained similarity (or correlation) between different countries in the virus spread process and helps identify the most influential countries that exhibit strong representativeness. More specifically, the technique extracts the graph Laplacian matrix $\mathbf{L}$ of the COVID-19 data in each period by first deriving $\mathbf{U}$ and then solving $\boldsymbol{\Lambda}$ efficiently using convex optimization techniques. By applying the extracted graph and centrality measures, influential countries significant in the study of the COVID-19 spread are identified.

7.3.1 Graph Topology Extraction

First, $\mathbf{L}$ is estimated and the graph structure $\mathcal{G}$ substantiates the COVID-19 data $\mathbf{X}$. By taking into account the band-limitedness of $\mathbf{X}$, the problem is formulated as

$$\min_{\mathbf{L},\mathbf{U}\in\mathbb{R}^{N\times N},\mathbf{S}\in\mathbb{R}^{N\times P}} \|\mathbf{X} - \mathbf{US}\|_F^2 + f(\mathbf{L},\mathbf{X}) \qquad (7.4a)$$

$$\text{s.t.} \quad \mathbf{U}^T\mathbf{U} = \mathbf{I}_N, \qquad (7.4b)$$

$$\mathbf{S} \in \mathcal{B}_K, \qquad (7.4c)$$

$$\mathbf{L} = \mathbf{U}\mathbf{\Lambda}\mathbf{U}^T, \mathbf{L} \in \mathbb{L}, \text{tr}(\mathbf{L}) = N, \qquad (7.4d)$$

$$\mathbf{u}_1 = \frac{1}{\sqrt{N}}\mathbf{1}. \qquad (7.4e)$$

The objective (7.4a) is composed of two terms. The first accounts for data consistency by penalizing any discrepancy between $\mathbf{US}$ and $\mathbf{X}$ using a quadratic loss. The second term provides a regularization function. According to [44] and [140], the function is set as

$$f(\mathbf{L},\mathbf{X}) = \text{tr}(\mathbf{X}^T\mathbf{LX}) + \alpha \left\|\text{vec}(\mathbf{L})\right\|_1.$$

Constraint (7.4b) guarantees that the matrix $\mathbf{U}$ is unitary, satisfying the decomposition in (7.1); constraint (7.4c) enforces that the GFT coefficient matrix $\mathbf{S}$ is K-block sparse; constraint (7.4d) ensures that $\mathbf{L}$ complies with the requirement of a legitimate graph Laplacian, and $\mathbb{L}$ contains all legitimate candidates for $\mathbf{L}$ [52]:

$$\mathbb{L} = \{\mathbf{L} \succeq \mathbf{0} | \mathbf{L1} = \mathbf{0}, L_{ij} = L_{ji} \leq 0, \forall i \neq j\}. \qquad (7.5)$$

According to $\mathbf{L1} = \mathbf{0}$ in (7.5), it is concluded that 0 is an eigenvalue of $\mathbf{L}$ and corresponds to the eigenvector $\mathbf{u}_1 = \frac{1}{\sqrt{N}}\mathbf{1}$, i.e., the first column of $\mathbf{U}$; see (7.4e).

Remark 5 *To address the non-convexity of (7.4) caused by the non-convex orthonormality constraint in (7.4b) and the sparsity constraint in (7.4c), (7.4) is decoupled and solved in two phases. Given the COVID-19 data $\mathbf{X}$, the GFT basis $\mathbf{U}$ is first estimated by minimizing $\|\mathbf{X} - \mathbf{US}\|_F^2$ subject to $\mathbf{U}^T\mathbf{U} = \mathbf{I}_N$, $\mathbf{S} \in \mathcal{B}_K$, and $\mathbf{u}_1 = \frac{1}{\sqrt{N}}\mathbf{1}$. In the second step, the eigenvalues $\mathbf{\Lambda}$ is estimated by minimizing the regularizer $\text{tr}(\mathbf{X}^T\mathbf{LX}) + \alpha \left\|\text{vec}(\mathbf{L})\right\|_1$ with the obtained $\mathbf{U}$.*

7.3.1.1 Extraction of Eigenvectors

Starting with the GFT basis, $\mathbf{U}$, provides a way to identify the intrinsic structure in the COVID-19 data that are related to the underlying pandemic network, even without the *a-priori* information of graph Laplacian $\mathbf{L}$. To estimate $\mathbf{U}$ from $\mathbf{X}$ satisfies the definition of GFT, i.e., $\mathbf{X} = \mathbf{U}\mathbf{S}$.

By utilizing the orthonormality property of $\mathbf{U}$ in (7.4b), it has $\|\mathbf{X} - \mathbf{US}\|_F^2 = \|\mathbf{U}^T\mathbf{X} - \mathbf{S}\|_F^2$. The first part of problem (7.4) is addressed, as given by [140, eq. 8]

$$\min_{\mathbf{U}\in\mathbb{R}^{N\times N}, \mathbf{S}\in\mathbb{R}^{N\times P}} \left\|\mathbf{U}^T\mathbf{X} - \mathbf{S}\right\|_F^2, \text{ s.t. (7.4b), (7.4c), (7.4e).} \qquad (7.6)$$

Despite the convex objective function, problem (7.6) is non-convex due to the orthonormality in (7.4b) and the sparsity in (7.4c). Since both $\mathbf{U}$ and $\mathbf{S}$ are unknown, (7.6) is reorganized as

$$\min_{\mathbf{U}\in\mathbb{R}^{N\times N}, \mathbf{S}\in\mathcal{B}_K} \sum_{i=1}^{N} \left\|\mathbf{u}_i^T\mathbf{X} - \mathbf{S}(i,:)\right\|_2^2, \text{ s.t. (7.4b), (7.4e),} \qquad (7.7)$$

which can be rewritten as

$$\min_{\mathbf{U}\in\mathbb{R}^{N\times N}} \left(\min_{\mathbf{S}\in\mathcal{B}_K} \sum_{i\in\mathcal{K}} \left\|\mathbf{u}_i^T\mathbf{X} - \mathbf{S}(i,:)\right\|_2^2 + \sum_{i\notin\mathcal{K}} \left\|\mathbf{u}_i^T\mathbf{X}\right\|_2^2 \right)$$

$$\text{s.t. (7.4b), (7.4e).} \qquad (7.8)$$

By closely analyzing the objective function of (7.8), it can be noticed that the optimal $\mathcal{K}$ comprises K most significant entries of $\{\|\mathbf{u}_i^T\mathbf{X}\|\}_i^N$, and satisfies

$$\mathbf{S}(i,:) = \begin{cases} \mathbf{u}_i^T\mathbf{X}, & \text{if } i \in \mathcal{K}; \\ \mathbf{0}, & \text{otherwise.} \end{cases} \qquad (7.9)$$

Therefore, the objective of (7.8) is reduced to only include the $(N-K)$ smallest entries of $\{\|\mathbf{u}_i^T\mathbf{X}\|\}_i^N$, after optimizing $\mathbf{S}$ to suppress $\sum_{i\in\mathcal{K}} \|\mathbf{u}_i^T\mathbf{X} - \mathbf{S}(i,:)\|_2^2$ using (7.9). To minimize this objective with respect to $\mathbf{S}$, the aim is to seek the optimal $\mathbf{U}$, represented as $\mathbf{U}^*$, in (7.6).

Substitute (7.9) into the objective of (7.8). Then, problem (7.6) can be written as

$$\mathbf{U}^* = \arg\min_{\mathbf{U}} \sum_{i\notin\mathcal{K}} \left\|\mathbf{u}_i^T\mathbf{X}\right\|_2^2 = \arg\min_{\mathbf{U}} \left\|\mathbf{U}_{\mathcal{K}^c}^T\mathbf{X}\right\|_F^2$$

$$= \arg\max_{\mathbf{U}} \left\|\mathbf{U}_{\mathcal{K}}^T\mathbf{X}\right\|_F^2, \qquad (7.10)$$

where $\mathcal{K}^c$ denotes the complementary set of $\mathcal{K}$, i.e., $\mathcal{K}^c = \mathcal{V} \setminus \mathcal{K}$; and the matrices $\mathbf{U}_{\mathcal{K}}$ and $\mathbf{U}_{\mathcal{K}^c}$ collate the column-vectors of $\mathbf{U}$ with indexes collected in $\mathcal{K}$ and $\mathcal{K}^c$, respectively.

Despite the non-convexity of (7.10), the goal of (7.10) is to identify the K-dimensional subspace in which the COVID-19 data $\mathbf{X}$ has the largest orthogonal projection; i.e.,

$$\arg\max_{\mathbf{U}} \left\|\mathbf{U}_{\mathcal{K}}^T\mathbf{X}\right\|_F^2 = \arg\max_{\mathbf{U}} \text{tr}\left(\mathbf{P}_{\mathbf{U}_{\mathcal{K}}}\mathbf{X}\mathbf{X}^T\right), \qquad (7.11)$$

where $\mathbf{P}_{\mathbf{U}_{\mathcal{K}}} = \mathbf{U}_{\mathcal{K}}\mathbf{U}_{\mathcal{K}}^T$ is the orthogonal projector onto the subspace spanned by $\mathbf{U}_{\mathcal{K}}$.

Using (7.11), the problem (7.6) is reformulated as

$$\mathbf{U}^* = \arg\max_{\mathbf{U}} \mathrm{tr}\left(\mathbf{P}_{\mathbf{U}_{\mathcal{K}}}\mathbf{X}\mathbf{X}^T\right), \quad \text{s.t. (7.4e).} \qquad (7.12)$$

Theorem 2 *By examining the two cases of* $\mathbf{u}_1 \notin \mathbf{U}_{\mathcal{K}}$ *and* $\mathbf{u}_1 \in \mathbf{U}_{\mathcal{K}}$, *the optimal solution to problem (7.6), denoted by* $\mathbf{U}^* = [\mathbf{U}_{\mathcal{K}}^*, \mathbf{U}_{\mathcal{K}^c}^*]$, *can be obtained as*

$$\mathbf{U}^* = \mathrm{eigen}\left[\left(\mathbf{I} - \mathbf{u}_1\mathbf{u}_1^T\right)\mathbf{X}\mathbf{X}^T\left(\mathbf{I} - \mathbf{u}_1\mathbf{u}_1^T\right)^T\right]. \qquad (7.13)$$

7.3.1.2 Extraction of Eigenvalues

Given the $\mathcal{K}$-band-limited COVID-19 data with the optimal $\mathbf{U}^*$ gained from (7.13), the graph Laplacian $\mathbf{L}$ is written as

$$\mathbf{L} = [\mathbf{U}_{\mathcal{K}}, \mathbf{U}_{\mathcal{K}^c}]\begin{bmatrix}\boldsymbol{\Lambda}_{\mathcal{K}} & \\ & \boldsymbol{\Lambda}_{\mathcal{K}^c}\end{bmatrix}[\mathbf{U}_{\mathcal{K}}, \mathbf{U}_{\mathcal{K}^c}]^T. \qquad (7.14)$$

where $\boldsymbol{\Lambda} = \begin{bmatrix}\boldsymbol{\Lambda}_{\mathcal{K}} & \\ & \boldsymbol{\Lambda}_{\mathcal{K}^c}\end{bmatrix}$. By plugging (7.14), $\mathrm{tr}(\mathbf{X}^T\mathbf{L}\mathbf{X})$ is written as

$$\begin{aligned}\mathrm{tr}(\mathbf{X}^T\mathbf{L}\mathbf{X}) &= \mathrm{tr}(\mathbf{X}^T(\mathbf{U}_{\mathcal{K}}\boldsymbol{\Lambda}_{\mathcal{K}}\mathbf{U}_{\mathcal{K}}^T)\mathbf{X} + \mathbf{X}^T(\mathbf{U}_{\mathcal{K}^c}\boldsymbol{\Lambda}_{\mathcal{K}^c}\mathbf{U}_{\mathcal{K}^c}^T)\mathbf{X}) \\ &= \mathrm{tr}(\mathbf{S}_{\mathcal{K}}^T\boldsymbol{\Lambda}_{\mathcal{K}}\mathbf{S}_{\mathcal{K}}).\end{aligned} \qquad (7.15)$$

Problem (7.4) becomes

$$\min_{\boldsymbol{\Lambda}_{\mathcal{K}}, \boldsymbol{\Lambda}_{\mathcal{K}^c}, \mathbf{L}} \mathrm{tr}(\mathbf{S}_{\mathcal{K}}^T\boldsymbol{\Lambda}_{\mathcal{K}}\mathbf{S}_{\mathcal{K}}) + \alpha\,\|\mathrm{vec}(\mathbf{L})\|_1$$

$$\begin{aligned}\text{s.t.} \quad &\mathbf{L} = [\mathbf{U}_{\mathcal{K}}, \mathbf{U}_{\mathcal{K}^c}]\begin{bmatrix}\boldsymbol{\Lambda}_{\mathcal{K}} & \\ & \boldsymbol{\Lambda}_{\mathcal{K}^c}\end{bmatrix}[\mathbf{U}_{\mathcal{K}}, \mathbf{U}_{\mathcal{K}^c}]^T, \\ &\boldsymbol{\Lambda}_{\mathcal{K}} \succeq 0, \boldsymbol{\Lambda}_{\mathcal{K}^c} \succeq 0, \\ &\mathbf{L}\mathbf{1} = 0, \\ &\mathrm{tr}(\mathbf{L}) = N, \\ &L_{ij} = L_{ji} \leq 0, \forall i \neq j.\end{aligned} \qquad (7.16)$$

Since its objective and constraints are convex or affine, problem (7.16) is convex and can be effectively addressed using by CVX toolboxes. With $\mathbf{U}$ and $\boldsymbol{\Lambda}$ obtained, it can obtain the graph Laplacian $\mathbf{L}$ underlying the European COVID-19 data using (7.1).

7.3.2 Influential Country Identify

Next, given the graph topology $\mathbf{L}$ underlying the COVID-19 data and indicating the propagation of the virus, the spread pattern of the four variants

Table 7.1: Topological characteristics of the learned complex networks.

Metric	Formula	Description
Degree centrality	$C_d(n_i) = \frac{\sum_1^j e_{ij}}{N-1}$	The number of edges directed toward node i.
Closeness centrality	$C_c(n_i) = \frac{N-1}{\sum_{i \neq j} d_{ij}}$	The average length of the shortest paths from node i to the rest of the nodes.
Betweenness centrality	$C_b(n_i) = \frac{\sum\limits_{i,j \neq v} \frac{\sigma_{ij}(v)}{\sigma_{ij}}}{(N-1)(N-2)}$	The frequency of a node serves as an intermediate relay along the shortest paths.
Average path length	$\frac{\sum_{i \neq j} d_{ij}}{N(N-1)}$	The average length of all the shortest paths in a graph.
Global efficiency	$\frac{N(N-1)}{\sum_{i \neq j} d_{ij}}$	The efficiency of information exchange between all node pairs.

among the European countries is estimated. As shown in Table 7.1, three node-level metrics, including degree centrality [152], closeness centrality [32], and betweenness centrality [16], are used to measure the influence of individual countries in the COVID-19 spread, where d_{ij} represents the shortest distance between nodes i and j in the extracted graph, σ_{ij} is the total amount of shortest paths between nodes i and j, and $\sigma_{ij}(v)$ denotes the number of these paths through node v.

- Degree centrality measures the number of connections a node has, helping identify the most connected nodes to the rest of the pandemic networks [152].

- Closeness centrality measures the inverse of the sum of the distances between a node and all other nodes in the network, which helps to identify nodes that are central and easily reachable within the network [32].

- Betweenness centrality quantifies the importance of a node in maintaining the shortest paths between other nodes in the network, helping to identify nodes that play a critical role in connecting different parts of the network [16].

The higher centrality a country has, the more influential it is and the more attention it deserves. In other words, the countries ranked high in terms of the centrality measures are likely to present important COVID-19 spread patterns.

Many other existing methods, such as node embeddings [186], DeepWalk [14], spectral clustering [156], and influence maximization [64], aimed to efficiently find influential nodes in large-scale graphs, e.g., social networks with thousands or even millions of nodes, often still based on the above classical centrality measures. Nevertheless, the graph considered consists of only $N = 44$ vertices (for 44 European countries). Computational complexity is less of a concern.

Two network-level metrics are also considered in Table 7.1, i.e., average path length [172] and global efficiency [145], to explore the spread of the pandemic.

- Average path length measures the average number of hops needed to get from one node to another node in the network [172]. A short average path length indicates a highly connected network, contributing to the fast spread of the pandemic [41].

- Global efficiency measures the average inverse shortest path length between all pairs of nodes, indicating how quickly the virus can spread [145]. A high global efficiency indicates a dense and well-connected network with fast virus propagation, while a low global efficiency indicates a fragmented and poorly connected network, deterring virus propagation.

7.4 Method Assessment and Results

In this section, First, the superiority of the proposed technique over existing approaches in graph learning accuracy of the COVID-19 data is experimentally validated. Then, the technique is used to conduct an in-depth analysis of the COVID-19 data, providing new insights into the pandemic's spread compared to existing techniques. The analysis is based on the open-access WHO dataset of daily identified COVID-19 cases in 44 European countries.

Apart from the proposed method, the state-of-the-art solutions are evaluated: Saboksayr's algorithm [135], Sardellitti's TV algorithm [140], Sardellitti's ESA algorithm [140], and Humbert's algorithm [80].

- Saboksyr' algorithm [86]: This is a scalable and time-efficient primal-dual algorithm that learns the topological structures of time series represented by the weighted adjacency matrices of graphs. However, this method has no explicit generative model for the observations. In other words, the model's accuracy may not be adequate for numerous real-world datasets that exhibit localized behaviors or piecewise smoothness.

- Sardellitti's TV graph learning algorithm [140]: The approach involves a two-step scheme: (*a*) learning the orthonormal and sparse transform of the data using AO, and (*b*) inferring the Laplacian from the sparsifying transform using convex optimization. The algorithm is reasonably computationally efficient by exploiting convex optimization techniques. However, the effectiveness of the overall process is compromised due to the AO-based approximation in the first step, which penalizes the fidelity of the orthogonal sparsifying transform.

- Sardellitti's ESA graph learning algorithm [140]: Different from Sardellitti's TV graph learning algorithm, this algorithm utilizes the information of the GFT coefficient matrix of the first step in the second step, where the graph Laplacian is recovered from the sparsifying transform and the GFT coefficients using convex optimization.

- Humbert's algorithm [80]: This is another AO-based algorithm that runs Riemannian manifold gradient descent and linear cone programs in an alternating fashion. However, only suboptimal solutions can be obtained using the AO method. The computational cost is also high.

Apart from the aforementioned advanced graph learning techniques, the proposed algorithm is also evaluated in comparison to the state-of-the-art GNN [169] when assessing the accuracy of the algorithm. The GNN consists of multiple hidden layers with 50 hidden units per layer. In the training stage, the input to the GNN includes the training data and the weighted correlation matrix of the training set. By contrast, the training set serves as the input for the graph learning algorithms.

7.4.1   Graph Learning-Based Analysis of COVID-19 Data

Fig. 7.2 provides the pandemic spread networks of 44 European countries over the four different periods obtained by the proposed algorithm, where the parameters of the algorithm are $K = 26$ and $\alpha = 1$ decided in the way delineated at the beginning of Section 7.4.2. The thickness of an edge measures the similarity of the COVID-19 spread between two countries. The virus spreads in the two countries are more likely to be related if the edge is thicker. The density of the edges indicates the extent to which COVID-19 spread among countries. It is observed in Fig. 7.2 that the virus spreads are increasingly related among the European countries from Period 1 to Period 4. Not only did the spread increase between the countries, but the virus spread increasingly widely across more countries.

To better illustrate the correlation of the COVID-19 spread between the European countries, Fig. 7.3 plots the weighted adjacency matrices of the graphs extracted from the COVID-19 data by the proposed algorithm. In the figure, the 44 European countries are sorted alphabetically from Albania to Ukraine along the x- and y-axes. The intensity of the color at each pixel stands for the extent of the correlation between the two countries associated with the pixel.

For example, the pixel corresponding to Greece and Norway is lighter than others in Fig. 7.3(a), indicating that Greece and Norway are highly correlated in Period 1. Likewise, Russia and Belarus are highly correlated in Period 2 in Fig. 7.3(b). Nevertheless, the number of light-colored pixels increases overall

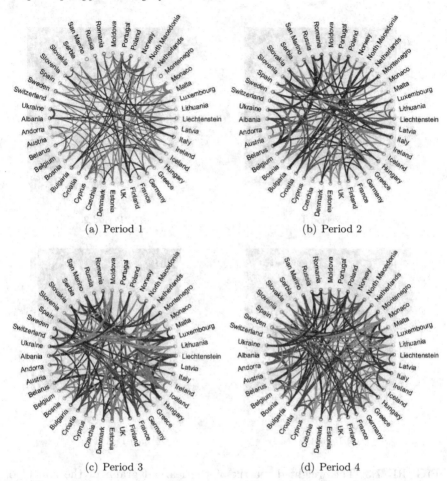

(a) Period 1

(b) Period 2

(c) Period 3

(d) Period 4

FIGURE 7.2: The learned graph of the COVID-19 spread in the 44 European countries during different periods.

in both Periods 3 and 4 in Figs. 7.3(c) and 7.3(d), indicating that the Delta and Omicron variants have higher and stronger propagation characteristics in Europe, which is consistent with the finding made in Fig. 7.2.

Figs. 7.4–7.7 visualize the top 5 countries that are identified to have been the most influential in the process of the COVID-19 virus spread in Europe, using the proposed approach based on the aforementioned three node-level metrics, i.e., degree centrality, closeness centrality, and betweenness centrality. A darker color indicates a country identified by more centrality measures to be among the top 5 most influential countries. For example, Czechia was influential during Period 1 in the sense of all three centrality measures. This makes sense since the different centrality measures are closely related in nature [130].

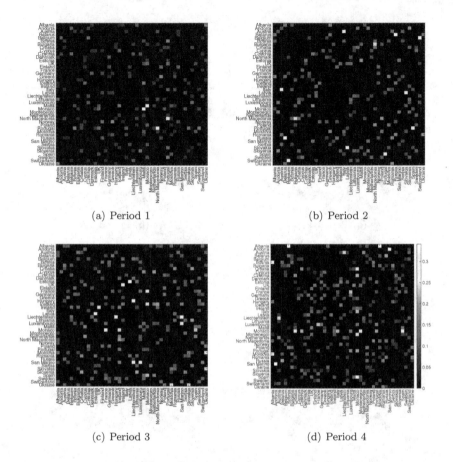

(a) Period 1   (b) Period 2

(c) Period 3   (d) Period 4

FIGURE 7.3: The weighted matrix of the learned graph of the spread of COVID-19 in 44 European countries during different periods.

It is obvious in Figs. 7.4–7.7 that the proposed algorithm identifies a different set of the most influential European countries in the COVID-19 spread, compared to the current advanced graph learning methods. Particularly, our new algorithm helps identify a small and concentrated set of influential countries in every period of COVID-19 spread; i.e., a country is more likely to be associated with multiple centrality measures. In other words, the influence of a country is more likely to be manifested through multiple measures. Here, the parameters of each method are separately tested and optimized, according to their individual settings.

Fig. 7.8 quantitatively evaluates how different the top 5 most influential countries are identified by the different algorithms. Specifically, the 15 most important countries identified using each of the considered algorithms based

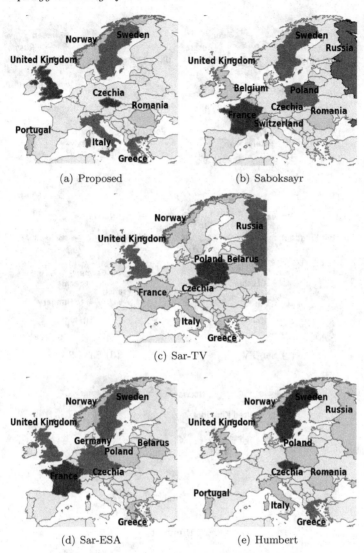

FIGURE 7.4: Influential countries identified during Period 1.

on the three centrality measures are vectorized. The similarity between the 15-element vectors produced by any two of the considered graph learning algorithms, measured by the cosine distance $\frac{\mathbf{V}_1^T \mathbf{V}_2}{|\mathbf{V}_1| \cdot |\mathbf{V}_2|}$, quantifies the similarity between the algorithms, where $\mathbf{V}_1$ and $\mathbf{V}_2$ are the two 15-element column vectors, and $|\cdot|$ stands for the norm.

As shown in Fig. 7.8, our new graph learning algorithm yields the highest similarity to Humbert's [80] in terms of their identified important countries

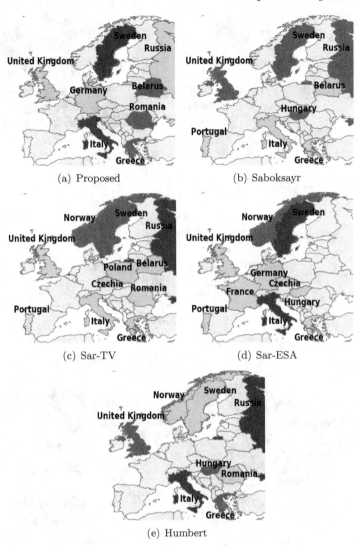

(a) Proposed

(b) Saboksayr

(c) Sar-TV

(d) Sar-ESA

(e) Humbert

FIGURE 7.5: Influential countries identified during Period 2.

(under three centrality measures), followed by Sar-ESA [140], Sar-TV [140], and Saboksayr's [135]. The similarities of the proposed algorithm to the existing algorithms are consistent with the graph learning (and reconstruction) accuracy of the algorithms, as will be shown in Fig. 7.12. Note that the ground truth regarding the most important countries is unavailable in practice. Given the best graph learning accuracy of the proposed algorithm and the consistent rankings between the accuracies and the similarities of the existing algorithms, it is reasonable to conclude that the countries identified by the

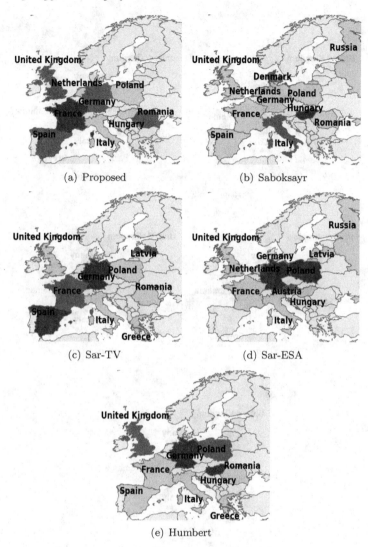

(a) Proposed

(b) Saboksayr

(c) Sar-TV

(d) Sar-ESA

(e) Humbert

FIGURE 7.6: Influential countries identified during Period 3.

proposed algorithm are more accurate and can contribute to more effective study and response to the pandemic.

Fig. 7.9 plots the average path length and global efficiency of the graph recovered by the proposed graph learning algorithm in the four periods of the COVID-19 pandemic. It is observed that the average path length decreases while the global efficiency increases during the four periods. The Omicron variant (i.e., Period 4) corresponds to the shortest average path length and the highest global efficiency, indicating that the Omicron variant has a higher

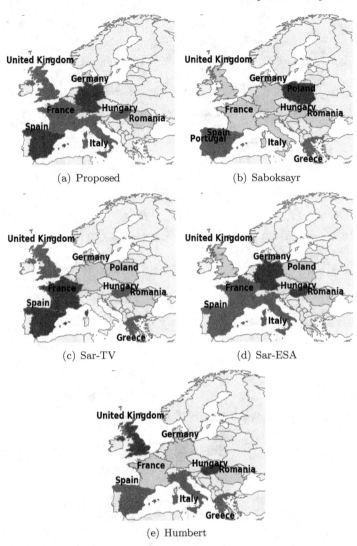

(a) Proposed   (b) Saboksayr

(c) Sar-TV   (d) Sar-ESA

(e) Humbert

FIGURE 7.7: Influential countries identified during Period 4.

level of global reachability and infectivity. In contrast, the original strains in the early stage of the pandemic, i.e., Period 1, have higher average path lengths and smaller global efficiencies. This is consistent with the finding in Figs. 7.2 and 7.3. The reason can be that during Period 1, the countries responded to the outbreak with stay-at-home or workplace closure, effectively slowing down the increase in confirmed cases.

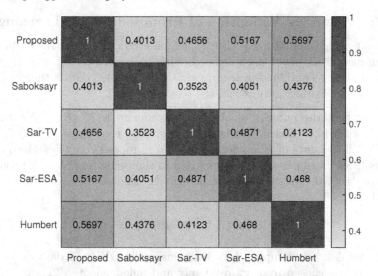

FIGURE 7.8: The correlation of different algorithms.

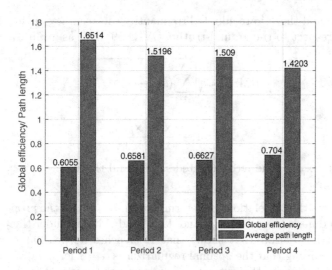

FIGURE 7.9: Average path length and global efficiency corresponding to different periods of COVID-19.

7.4.2 Accuracy Validation of Proposed Graph Learning

Without the ground truth of the graphs underlying the COVID-19 data, the learning accuracy of the proposed algorithm is assessed by obfuscating part of the data. The reconstruction accuracy of this obfuscated data is then evaluated based on the learned graphs and the remaining data.

Suppose that the number of observable countries is K ($K \leq N$), i.e., the signal bandwidth. Based on the inferred graphs, e.g., those in Fig. 7.3, and the COVID-19 data of K randomly selected European countries, the quantity of identified cases per million population in the remaining $(N - K)$ countries is reconstructed. The recovered graph signals, denoted by $\hat{\mathbf{x}}_p$, can be obtained as [21]

$$\hat{\mathbf{x}}_P = \mathbf{U}_{\mathcal{K}} \mathbf{U}_{\mathcal{K}}^T \boldsymbol{\Psi}^T \boldsymbol{\Psi} \mathbf{D}^2 \boldsymbol{\Psi}^T \mathbf{y}_P, \qquad (7.17)$$

where $\mathbf{y}_p \in \mathbb{R}^K$ is sampled $K \times 1$-dimensional COVID-19 data on the p-th day, which is chosen from $\mathbf{x}_p$ randomly and independently [21]. $\boldsymbol{\Psi} \in \mathbb{R}^{K \times N}$ stands for a sampling operator. $\boldsymbol{\Psi}_{ij} = 1$ if $j = \mathcal{K}_i$; and 0, otherwise. Here, $\mathcal{K}_i$ is the i-th element of $\mathcal{K}$, indicating the i-th of the $K = |\mathcal{K}|$ European countries with COVID-19 data available. $\mathbf{D} \in \mathbb{R}^{N \times N}$ is a diagonal rescaling matrix with $D_{ii} = 1/\sqrt{K\pi_i}$ and π_i being the probability of choosing the i-th $K \times N$-dimensional sample of the K countries on the p-th day of the considered period. Since a uniform sampling process is considered, the sampling metric for every node is $\pi_i = 1/N$.

The RMSE and the R^2 are adopted to quantify the accuracy of the recovered data with respect to the ground-truth COVID-19 data, as given by

$$\text{RMSE} = \sqrt{\sum_{i=1}^{N} (\hat{x}_{pi} - x_{pi})^2 / N}; \qquad (7.18)$$

$$R^2 = 1 - \frac{\|\hat{\mathbf{x}}_p - \mathbf{x}_p\|_2^2}{\|\hat{\mathbf{x}}_p - \bar{\mathbf{x}}_p\|_2^2}. \qquad (7.19)$$

Here, $\hat{\mathbf{x}}_p$ and $\bar{\mathbf{x}}_p$ are the reconstructed signals and the average of the ground-truths of $\mathbf{x}_p$.

Fig. 7.10 plots the correlations of determination, i.e., R^2, of the proposed algorithm with different regularizer α and data bandwidth K under the pandemic network during Period 1. It shows that R^2 reaches its peak at $\alpha = 1$ and $K = 26$; indicating that the optimal regularizer is $\alpha = 1$ for a data bandwidth of $K = 26$. Similarly, the optimal values of α for Periods 2 to 4 are determined. Fig. 7.11 shows the R^2 of the considered graph learning algorithms in four different periods, where $K = 26$. The proposed algorithm obtains the largest R^2. For example, the improvements of the algorithm are about 29.36%, 27.71%, 12.46%, and 11.11%, compared to Saboksayr's [135], Sar-TV [140],

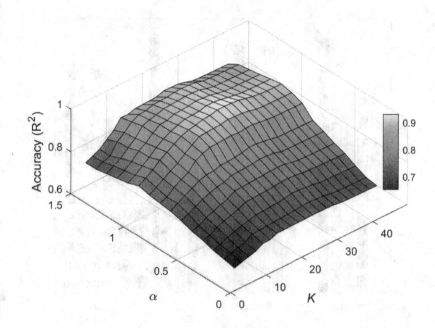

FIGURE 7.10: The accuracy vs. bandwidth K and α in Period 1.

Sar-ESA [140], and Humbert's [80], respectively. To ensure a fair comparison, the parameters are individually tested and optimized for each benchmark in these figures.

Fig. 7.12(a) shows the RMSE of the considered algorithms with the increase in the signal bandwidth K. It shows that under all the considered algorithms, the RMSEs decrease quickly with the growth of K and then converge to constant values. Our proposed algorithm has the smallest RMSE under all values of K. It has the minimum RMSE around 0.23 at $K = 26$ and achieves performance improvements by about 60.87%, 43.48%, 34.78%, and 33.33%, compared to Saboksayr's [135], Sar-TV [140], Sar-ESA [140], and Humbert's [80], respectively. Fig. 7.12(b) plots the cumulative distribution function (CDF) of the errors undergone by the considered algorithms. As depicted, our new algorithm exhibits significantly lower estimation errors compared to the other algorithms. In particular, over 80% of the estimation errors are smaller than 0.2 cases per million population under our algorithm. By contrast, 38.3%, 48.4%, 59.5%, and 64.6% of the estimation errors are smaller than 0.2 cases per million population under Saboksayr's [135], Sar-TV [140], Sar-ESA [140], and Humbert's [80], respectively.

Next, the accuracy (R^2) of the considered graph learning algorithms is assessed by predicting future missing data based on the graph topologies extracted from

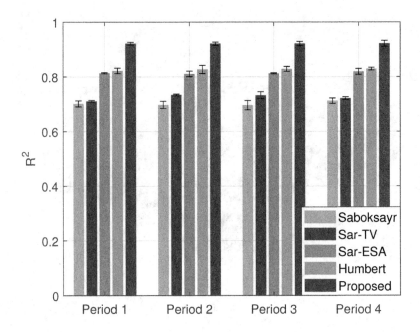

FIGURE 7.11: Efficiency of reconstruction of different methods upon four periods when $K = 26$.

past data. In addition to the graph learning techniques, the state-of-the-art GNN [169] is also considered. The COVID-19 dataset of each period is divided into a training set (e.g., the first 80% of the dataset) and a test set (e.g., the remaining 20% of the dataset). In the training phase, the graph learning algorithms extract the graph topology of the training set. In the test phase, the test data of Ukraine is assumed to be missing and is predicted based on the graph topologies extracted from the training set and the available test data of the other countries. By adjusting the ratio between the training and test sets, the robustness of the algorithms to the small training set is demonstrated.

As shown in Figs. 7.13(a)–7.13(d), the graph learning methods, including our proposed algorithm, outperform the GNN under different ratios between the training and test sets. When the training set is set to 80%, and the testing set is 20%, our algorithm achieves the highest R^2 values with the improvements of about 70.49%, 75.85%, 70.99%, and 68.11% in the four periods, compared to the GNN. Notice that the R^2 value of the GNN can yield negative values, particularly in cases where the training set is limited. This is the case when even the mean of the data can provide a better fit to the data than the fitted function, e.g., the GNN, when the training set is limited, i.e., 20%.

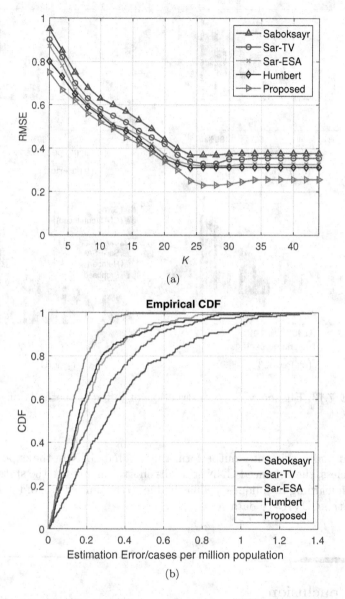

FIGURE 7.12: (a) The RMSE vs. the bandwidth K. (b) The CDFs of estimation error under different graph learning methods.

On the other hand, our new algorithm can enhance the state-of-the-art GNN by providing more accurate graph topologies, compared to a direct calculation of adjacency matrices (as done in the GNN [169]). By inputting the weighted adjacency matrices of the graphs learned by the algorithm, the GNN can be

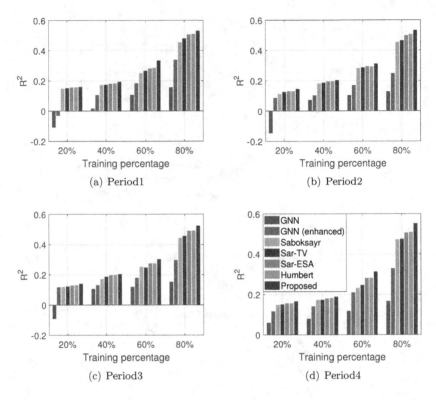

FIGURE 7.13: Efficiency of reconstruction of different methods of different periods.

enhanced and consistently outperform the original GNN in the experiments. Nevertheless, the enhanced GNN still falls short compared to the state-of-the-art graph learning techniques, primarily due to the limited size of the training set, i.e., the COVID-19 data set.

7.5   Conclusion

In this chapter, a new graph-learning technique is presented to analyze the evolution of the COVID-19 pandemic and reveal the underlying relationship and spreading pattern among different countries. The new technique estimates the graph Laplacian of the COVID-19 data by first deriving the closed-form expression for its eigenvectors and then estimating its eigenvalues with convex optimization. Based on the COVID-19 data, the accuracy of the estimated graph Laplacian was shown to outperform the existing approaches by 33.3%

in RMSE and 11.11% in the correlation of determination. The new technique helped identify a different set of the most influential and representative European countries, in contrast to the previous techniques. Given the superior accuracy of the algorithm, the set of identified influential countries is expected to be sensible and deserves dedicated research efforts to help understand the COVID-19 spread.

8

Preserving the Privacy of Latent Information for Graph-Structured Data

Latent graph structure and stimulus of graph-structured data contain critical private information, such as brain disorders in functional magnetic resonance imaging data, and can be exploited to identify individuals. It is critical to perturb the latent information while maintaining the utility of the data, which, unfortunately, has never been addressed. This paper presents a novel approach to obfuscating the latent information and maximizing the utility. Specifically, the analysis begins with the GFT basis, which captures the latent graph structures, and the latent stimuli, which are the spectral-domain inputs to these latent graphs. Next, a new multi-objective problem is formulated and decoupled to alternately obfuscate the GFT basis and stimuli. The GFT basis is obfuscated using a combination of difference-of-convex (DC) programming and Stiefel manifold gradient descent. The DC programming and gradient descent are employed to perturb the spectral-domain stimuli. Experiments conducted on an attention-deficit hyperactivity disorder dataset demonstrate that this approach can substantially outperform its differential privacy-based benchmark in the face of the latest graph inference attacks.

8.1   Introduction

Graph data, characterized by latent graph structures, plays a crucial role in various fields, including physics, [106, 94], biology, transportation [35, 1, 159, 33], energy [96], engineering [176, 134, 34], and social science [111, 24]. Illustrative instances of such data encompass brain signals like Electroencephalography signals [138], and blood-oxygen-level-dependent time series derived from fMRI on the brain [61]. Moreover, social network data from platforms like Facebook, Twitter, and WeChat offer further examples, revealing social graphs and significant volumes of potentially sensitive and private user information [83, 158].

Privacy is a significant concern for some graph-structured data [114], for example, brain network data obtained by fMRI [61]. The graph-structured brain data can be held by the Department of Neurology in a hospital, and shared with and used by other departments or clinics for big data analytics (e.g., detecting or modeling changes in blood flow that occur with brain activity) or educational purposes.

On the one hand, the latent graph structures of brain network data could expose personal health conditions, such as ADHD [56] and AD [59], under graph interference attacks (based on graph learning techniques [142, 140, 80]). The health conditions could be exploited to reveal the identities of the patients [141, 127]. The latent stimuli of the graph-structured data, which are the input to the latent graphs and derive the output of the graph-structured data observed, are also part of the private information [116]. The bandwidth and waveform shapes of the stimuli can be used to identify individuals.

On the other hand, minimizing the perturbation on the observed graph-structured data, e.g., brain network data, helps retain the utility of the data, e.g., for measuring and modeling blood flows during brain activities, evaluating the effects of stroke, and examining functioning areas of the brain under different tasks. In this sense, it is critical to protect the privacy of the graph-structured data by obfuscating the latent graph structures and stimuli underlying the graph-structured data, while minimizing changes or perturbations to the graph-structured data to best maintain its utility. However, preserving the privacy of the graph data, more explicitly, the privacy of the latent graph structures and stimulus underlying the graph data, has never been addressed in the literature.

This chapter presents a new approach to preserving the privacy of the latent graph structures and stimuli of graph data while maximizing the utility of the graph data. The approach is important to defend against graph inference attacks, and can have extensive applications to protect personal health records (e.g., brain data), finance transactions, and many other graph-structured data.

- A new multi-objective problem is formulated to preserve the privacy of the latent graph structures and stimuli of graph data and maintain the utility of the data. The problem has never been addressed. The restrictiveness of DP in solving the problem is revealed.

- The analytical expressions for the GFT basis that captures the latent graph structures are derived, and the latent stimuli are the spectral-domain inputs to the latent graphs.

- The new multi-objective problem is decoupled to alternately obfuscate the GFT basis and stimulus against the expressions derived. The DC programming and Stiefel manifold gradient descent are orchestrated to perturb the GFT basis efficiently. The DC programming and gradient descent are employed to perturb the latent stimuli.

Extensive experiments are conducted on synthetic graph data generated under the Random Geometric model, and the real-world ADHD dataset. The new approach is demonstrated to effectively protect the privacy of the latent graph structures and stimuli of graph data, while maintaining the utility of the data. The approach can substantially outperform its DP-based benchmark in the face of graph inference attacks based on the latest graph learning techniques.

8.2 System Model and Problem Statement

In this section, the system model and problem statement are described.

8.2.1 System Model

Let $\mathbf{Y} \in \mathbb{R}^{N \times M}$ denote the observed data with a latent graph structure $\mathcal{G}(\mathcal{V}, \mathcal{E})$. N is the count of vertices on the latent graph. A $1 \times M$-dimensional time series is associated with each vertex in the observed data $\mathbf{Y}$. $\mathcal{V} = \{1, \cdots, N\}$ is the set of vertices, and $\mathcal{E} \subseteq \mathcal{V} \times \mathcal{V}$ is the set of edges. The topological structure of $\mathcal{G}$ can be captured by a weighted and undirected adjacency matrix $\mathbf{W} \in \mathbb{R}^{N \times N}$. The adjacency matrix $\mathbf{W}$ collects all the edges with $W_{ij} = W_{ji} \neq 0, \forall (i, j) \in \mathcal{E}$.

Let $\mathbf{D} \triangleq \text{diag}(\mathbf{W1})$ define the degree matrix containing the node degrees at its diagonal. Also, suppose that each node is connected to at least one other node, ensuring no isolated nodes are in the graph. In other words, none of the diagonal elements is zero in $\mathbf{D}$. Then, according to [52], the combinatorial graph Laplacian of $\mathcal{G}$ is defined as

$$\mathbf{L} = \mathbf{D} - \mathbf{W}, \tag{8.1}$$

which is a semi-definite matrix with positive elements along its main diagonal and non-positive elements anywhere else [52]. The eigenvectors of $\mathbf{L}$, denoted by $\mathbf{U}^* \in \mathbb{R}^{N \times N}$, make up the so-called GFT basis [23], which captures the latent graph structure or topology of the graph-structured data $\mathbf{Y}$ [52].

Apart from the latent graph structure, the observed graph-structured data $\mathbf{Y}$ also contains a latent stimulus in the spectral domain. The latent spectral-domain representation (or, in other words, the latent spectral-domain stimulus) of $\mathbf{Y}$ is represented by $\mathbf{S}^* \in \mathbb{R}^{N \times M}$. It projects $\mathbf{Y}$ on the spectral-domain subspace spanned by $\mathbf{U}^*$:

$$\mathbf{Y} = \mathbf{U}^* \mathbf{S}^*, \tag{8.2}$$

where $\mathbf{S}^*$ can exhibit some level of sparsity. Clearly, $\mathbf{S}^*$ relies on both $\mathbf{Y}$ and the latent graph structure characterized by the GFT basis, $\mathbf{U}^*$. Therefore, $\mathbf{S}^*$ and $\mathbf{U}^*$ need to be jointly estimated, which, however, is non-trivial [140].

As reported in [140], [143], and [91], the observed graph-structured data can often be sparse in the canonical domain. Let K denote the bandwidth of $\mathbf{S}^*$ in the spectral domain, $K \in \{1, \cdots, N\}$, and $\mathcal{B}_K$ collect all K-block sparse matrices [12]; i.e.,

$$\mathcal{B}_K \triangleq \{\mathbf{S}^* \in \mathbb{R}^{N \times M}, \mathbf{S}^*(i,:) = \mathbf{0}, \forall i \notin \mathcal{K} \in \mathcal{V}, |\mathcal{K}| = K\}, \tag{8.3}$$

where $\mathbf{S}^*(i,:)$ stands for the i-th row of $\mathbf{S}^*$, while $\mathcal{K} \in \mathcal{V}$ has the cardinality of K.

8.2.2 Problem Statement

In this chapter, the goal is to protect the privacy of the latent information underlying graph-structured data (e.g., the graph structure and the stimulus underlying the observed graph-structured data) while minimizing the perturbations on the observed graph-structured data to maintain the utility of the data. Let $\mathbf{U}^*$ and $\mathbf{S}^*$ denote the latent GFT basis and stimulus, respectively, and $\mathbf{U}$ and $\mathbf{S}$ denote their respective obfuscated versions.

The utility is measured by the difference between the observed graph data and their corresponding perturbed versions, i.e., $\|\mathbf{Y} - \mathbf{US}\|_F$. The privacy is measured by the difference between the latent graph structures of the graph data and their corresponding perturbed version, i.e., $\|\mathbf{U} - \mathbf{U}^*\|_F$; and by the difference between the latent stimuli of the graph data and their corresponding perturbed version, i.e., $\|\mathbf{S} - \mathbf{S}^*\|_F$.

The considered problem is formulated as

$$\max_{\mathbf{U} \in \mathbb{R}^{N \times N}, \mathbf{S} \in \mathbb{R}^{N \times M}} \|\mathbf{U} - \mathbf{U}^*\|_F^2 + \beta \|\mathbf{S} - \mathbf{S}^*\|_F^2 \tag{8.4a}$$

$$\text{and} \quad \min_{\mathbf{U} \in \mathbb{R}^{N \times N}, \mathbf{S} \in \mathbb{R}^{N \times M}} \|\mathbf{US} - \mathbf{Y}\|_F^2, \tag{8.4b}$$

$$\text{s.t.} \quad \mathbf{U}^T \mathbf{U} = \mathbf{I}_N, \tag{8.4c}$$

$$\mathbf{S} \in \mathcal{B}_K, \tag{8.4d}$$

$$\mathbf{u}_1 = 1/\sqrt{N} \ \mathbf{1}. \tag{8.4e}$$

Here, (8.4a) aims to prevent the leakage of the private information about the latent graph structure $\mathbf{U}^*$ and the latent stimulus $\mathbf{S}^*$ by maximizing their difference from their respective obfuscated versions. β is an adjustable hyperparameter to fine-tune the priority of the two terms in (8.4a). The second objective (8.4b) indicates the perturbed version of $\mathbf{Y}$ needs to be close to the original observation $\mathbf{Y}$ to maintain its utility.

Constraint (8.4c) is due to the orthonormal nature of $\mathbf{U}$. Constraint (8.4d) specify the sparsity of $\mathbf{S}$. Constraint (8.4e) provides the necessary condition of a valid graph Laplacian $\mathbf{L}$; i.e., one eigenvalue of $\mathbf{L}$ must be 0, and the corresponding eigenvector is $\mathbf{u}_1$, which is a column of $\mathbf{U}$ [140].

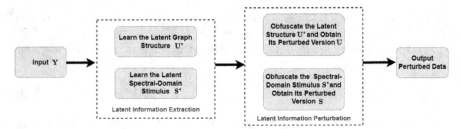

FIGURE 8.1: The flow diagram of the proposed obfuscation algorithm to perturb the latent information of graph-structured data.

8.3 Proposed Obfuscation for Graph Data

Fig. 8.1 depicts the diagram of the proposed obfuscation algorithm for graph-structured data, where there are two components: The extraction of the latent information of graph-structured data, and the perturbation of the latent information.

8.3.1 Latent Information Extraction

To preserve the privacy of the latent information $\mathbf{U}^*$ and $\mathbf{S}^*$, $\mathbf{U}^*$ and $\mathbf{S}^*$ based on the observed graph-structured data $\mathbf{Y}$ are estimated. Provided $K \in [1, N]$, the problem is casted as

$$(\mathbf{U}^*, \mathbf{S}^*) = \arg\min_{\mathbf{U}, \mathbf{S}} \|\mathbf{Y} - \mathbf{U}\mathbf{S}\|_F^2 \tag{8.5}$$

$$\text{s.t.}\quad (8.4\text{c}),\ (8.4\text{d}),\ (8.4\text{e}),$$

where, with a little abuse of notations, $\mathbf{U}$ and $\mathbf{S}$ are defined to be the estimation variables. The closed-form solution to problem (8.5) can be given in the following proposition.

Proposition 1 *Given $K \in [1, N]$, the solution to problem (8.5), $\mathbf{U}^*$ and $\mathbf{S}^*$, is uniquely given in a closed form by*

$$\mathbf{U}^* = \mathrm{Eig}\left[\left(\mathbf{I} - \mathbf{u}_1 \mathbf{u}_1^T\right) \mathbf{Y}\mathbf{Y}^T \left(\mathbf{I} - \mathbf{u}_1 \mathbf{u}_1^T\right)^T\right]; \tag{8.6}$$

$$\mathbf{S}^* = (\mathbf{U}^*)^T \mathbf{Y}, \tag{8.7}$$

where $\mathrm{Eig}[\mathbf{Z}]$ gives the eigenvectors of $\mathbf{Z}$.

Remark 6 *While the closed-form solution to problem (8.5), i.e., $\mathbf{U}^*$ and $\mathbf{S}^*$, require the knowledge of K, the solution given in (8.6) and (8.7) is suitable for any $K \in [1, N]$. For this reason, $K = N$ is configured to access the entire* spectrum of the stimulus of $\mathbf{Y}$. Subsequently, the norms of all rows in $\mathbf{S}^*$, i.e., $||\mathbf{S}^*(i,:)||$, are evaluated. K is typically determined by counting rows yielding larger norms than a predefined threshold.

8.3.2 Perturbation for Graph Privacy Preservation

Recall that $\mathbf{U}$ and $\mathbf{S}$ are the perturbed versions of $\mathbf{U}^*$ and $\mathbf{S}^*$ in (8.4), respectively. As stated in (8.4), $||\mathbf{U} - \mathbf{U}^*||_F$ and $||\mathbf{S} - \mathbf{S}^*||_F$ are maximized, i.e., the perturbations of the latent graph $\mathbf{U}^*$ and stimulus $\mathbf{S}^*$, while minimizing $||\mathbf{US} - \mathbf{Y}||_F$, i.e., the utility loss of the observed graph-structured data $\mathbf{Y}$.

Given $\mathbf{Y}$, $\mathbf{U}^*$, and $\mathbf{S}^*$, problem (8.4) can be recast to improve its mathematical tractability, as given by

$$\max_{\mathbf{S} \in \mathbb{R}^{N \times M}, \mathbf{U} \in \mathbb{R}^{N \times N}} \frac{||\mathbf{U} - \mathbf{U}^*||_F^2 + \beta ||\mathbf{S} - \mathbf{S}^*||_F^2}{||\mathbf{US} - \mathbf{Y}||_F^2} \tag{8.8}$$

$$\text{s.t.} \quad (8.4c), \ (8.4e).$$

Despite the conciseness of problem (8.8), it is challenging to obtain the optimal solution due to the non-convexity of objective function and the non-convexity and restrictive equality constraint in (8.4c). Problem (8.8) is addressed by using fractional programming [144]. By introducing an auxiliary variable α, problem (8.8) can be reformulated as

$$\max_{\mathbf{U},\mathbf{S}} \ ||\mathbf{U} - \mathbf{U}^*||_F^2 + \beta ||\mathbf{S} - \mathbf{S}^*||_F^2 - \alpha ||\mathbf{US} - \mathbf{Y}||_F^2 \tag{8.9}$$

$$\text{s.t.} \quad (8.4c), \ (8.4e).$$

In essence, the objective of problem (8.8) is to maximize α. Next, problem (8.9) is solved by optimizing α, $\mathbf{U}$, and $\mathbf{S}$ in an alternating manner, as delineated in the following.

8.3.2.1 Obfuscate Latent Graph Structure U

Given fixed α and $\mathbf{S}$ in the $(i-1)$-th AO iteration, i.e., $\alpha^{(i-1)}$ and $\mathbf{S}^{(i-1)}$ $(i = 1, 2, \cdots)$, in the i-th iteration, problem (8.9) is reduced to

$$\min_{\mathbf{U}} \ \alpha^{(i-1)} \left\| \mathbf{US}^{(i-1)} - \mathbf{Y} \right\|_F^2 - ||\mathbf{U} - \mathbf{U}^*||_F^2 \tag{8.10}$$

$$\text{s.t.} \quad (8.4c), \ (8.4e).$$

Consider the constraint (8.4e), $\mathbf{U} = [\mathbf{u}_1, \mathbf{V}]$ is rewritten, where $\mathbf{V} \in \mathbb{R}^{N \times (N-1)}$ collects the rest of the columns-vectors of $\mathbf{U}$ except the first column $\mathbf{u}_1$. Likewise, $\mathbf{S} = [\mathbf{s}_1, \mathbf{H}]^T$, where $\mathbf{s}_1$ is the first row-vector of $\mathbf{S}$ and $\mathbf{H} \in \mathbb{R}^{(N-1) \times M}$ collects the rest of the row-vectors of $\mathbf{S}$. Then, problem (8.10) can be rewritten as

$$\min_{\mathbf{V}} \; \alpha^{(i-1)} \left\| \mathbf{V}\mathbf{H}^{(i-1)} - \mathbf{Y} \right\|_F^2 - \| \mathbf{V} - \mathbf{V}^* \|_F^2 \qquad (8.11a)$$

$$\text{s.t.} \quad \mathbf{V}^T\mathbf{V} = \mathbf{I}_{N-1}, \qquad (8.11b)$$

$$\mathbf{u}_1^T\mathbf{V} = \mathbf{0}, \qquad (8.11c)$$

where $\mathbf{U}^* = [\mathbf{u}_1, \mathbf{V}^*]$ with $\mathbf{V}^* \in \mathbb{R}^{N \times (N-1)}$ collecting the rest of the column-vectors of $\mathbf{U}^*$ apart from $\mathbf{u}_1$.

The objective function in (8.11) has the form of DC program:

$$\min_{\mathbf{V}} \; g(\mathbf{V}) - h(\mathbf{V}), \qquad (8.12)$$

where $g(\mathbf{V}) = \alpha^{(i-1)} \left\| \mathbf{V}\mathbf{H}^{(i-1)} - \mathbf{Y} \right\|_F^2$ and $h(\mathbf{V}) = \| \mathbf{V} - \mathbf{V}^* \|_F^2$, both of which are convex functions.

By using the difference of convex algorithm (DCA) [65], problem (8.11) can be reformulated to a strongly convex optimization problem. At the l-th iteration of the DCA ($l = 1, 2, \cdots$), it has

$$\min_{\mathbf{V}} \; \alpha^{(i-1)} \left\| \mathbf{V}\mathbf{H}^{(i-1)} - \mathbf{Y} \right\|_F^2 - \langle \mathbf{V}, \nabla h(\mathbf{V}^{l-1}) \rangle \qquad (8.13a)$$

$$\text{s.t.} \quad \mathbf{V}^T\mathbf{V} = \mathbf{I}_{N-1}, \qquad (8.13b)$$

$$\mathbf{u}_1^T\mathbf{V} = \mathbf{0}, \qquad (8.13c)$$

where $\nabla h(\mathbf{V}^{l-1})$ is a gradient of $h(\mathbf{V})$ at $\mathbf{V}^{l-1}$, the local point obtained in the $(l-1)$-th iteration of the DCA.

The objective function (8.13a) is convex. Consider the orthonormal vector variables in constraint (8.13b); or in other words, the feasible solution region is on the Stiefel manifold. Problem (8.13) is convex on the Stiefel manifold and can be uniquely addressed by utilizing the Stiefel manifold gradient descent [129] with the details below.

The Lagrangian of (8.13) is denoted as

$$\mathcal{L}(\mathbf{V}, \boldsymbol{\Psi}, \boldsymbol{\Phi}) = \alpha^{(i-1)} \left\| \mathbf{V}\mathbf{H}^{(i-1)} - \mathbf{Y} \right\|_F^2 - \langle \mathbf{V}, \nabla h(\mathbf{V}^{l-1}) \rangle$$
$$- \frac{1}{2}\mathrm{tr}\left(\boldsymbol{\Psi}^T (\mathbf{V}^T\mathbf{V} - \mathbf{I}_{N-1}) \right) - \frac{1}{2}\mathrm{tr}\left(\boldsymbol{\Phi}^T (\mathbf{u}_1^T\mathbf{V}) \right), \qquad (8.14)$$

where $\boldsymbol{\Psi}$ and $\boldsymbol{\Phi}$ are the Lagrangian multipliers corresponding to (8.13b) and (8.13c), respectively.

Applying the Karush-Kuhn-Tucker (KKT) conditions, the Lagrangian function regarding $\mathbf{V}$ is differentiated and equates to zero:

$$\nabla_{\mathbf{V}}\mathcal{L}(\mathbf{V}, \boldsymbol{\Psi}, \boldsymbol{\Phi}) = \nabla \mathcal{F}(\mathbf{V}) - \mathbf{V}\boldsymbol{\Psi} - \frac{1}{2}\mathbf{u}_1\boldsymbol{\Phi} = \mathbf{0}, \qquad (8.15)$$

where $\mathcal{F}(\mathbf{V}) = \alpha^{(i-1)} \left\| \mathbf{V}\mathbf{H}^{(i-1)} - \mathbf{Y} \right\|_F^2 - \langle \mathbf{U}, \nabla h(\mathbf{V}^{l-1}) \rangle$ is defined for the brevity of notation, and $\nabla \mathcal{F}(\mathbf{V})$ is the gradient of $\mathcal{F}(\mathbf{V})$.

By left multiplying $\mathbf{V}^T$ on both sides of (8.15), it has

$$\mathbf{V}^T \nabla \mathcal{F}(\mathbf{V}) - \mathbf{V}^T \mathbf{V}\mathbf{\Psi} = \mathbf{0}. \tag{8.16}$$

(8.13b) is substituted into (8.16) and then (8.16) is reorganized, yielding

$$\mathbf{\Psi} = \mathbf{V}^T \nabla \mathcal{F}(\mathbf{V}). \tag{8.17}$$

By left multiplying $\mathbf{u}_1^T$ on both sides of (8.15), it has

$$\mathbf{u}_1^T \nabla \mathcal{F}(\mathbf{V}) - \frac{1}{2}\mathbf{u}_1^T \mathbf{u}_1 \mathbf{\Phi} = \mathbf{0}. \tag{8.18}$$

(8.13c) is substituted into (8.18) and then (8.18) is reorganized, yielding

$$\mathbf{\Phi} = 2\mathbf{u}_1^T \nabla \mathcal{F}(\mathbf{V}). \tag{8.19}$$

Since the constraint $\mathbf{V}^T\mathbf{V}$ is symmetric, $\mathbf{\Psi}$ is asymmetric and therefore $\mathbf{\Psi} = \nabla \mathcal{F}(\mathbf{V})^T \mathbf{V}$. By substituting (8.17) and (8.19) into (8.15), the gradient in (8.15) can be reformulated to

$$\nabla_{\mathbf{V}}\mathcal{L}(\mathbf{V}, \mathbf{\Psi}, \mathbf{\Phi}) = \nabla_{\mathbf{V}}\mathcal{L}(\mathbf{V}) \tag{8.20a}$$
$$= \nabla \mathcal{F}(\mathbf{V}) - \mathbf{V}\nabla \mathcal{F}(\mathbf{V})^T \mathbf{V} - \mathbf{u}_1 \mathbf{u}_1^T \nabla \mathcal{F}(\mathbf{V}). \tag{8.20b}$$

By running the Stiefel manifold gradient descent, problem (8.13) can be solved by iteratively updating the gradient of the Lagrange function $\nabla_{\mathbf{V}}\mathcal{L}(\mathbf{V})$ with (8.20) and $\mathbf{V}$ with the right-hand scaled gradient projection method [129]:

$$\mathbf{V} \leftarrow \pi(\mathbf{V}' - \tau_{\mathbf{V}}\nabla_{\mathbf{V}}\mathcal{L}(\mathbf{V})\mathcal{A}(\mu, \tau)), \tag{8.21}$$

where $\pi(\cdot)$ is the projection operator, i.e., $\pi(\mathbf{X}) = \mathbf{Q}\mathbf{I}\mathbf{P}^T$ if $\mathbf{X} = \mathbf{Q}\mathbf{\Sigma}\mathbf{P}^T$ by singular value decomposition (SVD) [129]; $\mathcal{A}(\mu, \tau_{\mathbf{V}})$ is a scaling matrix with $\mu \in (0, 1)$, i.e.,

$$\mathcal{A}(\mu, \tau_{\mathbf{V}})) = (\mathbf{I}_{N-1} + \mu\tau_{\mathbf{V}}\mathbf{V}^T\nabla_{\mathbf{V}}\mathcal{L}(\mathbf{V}))^{-1}, \tag{8.22}$$

and $\tau_{\mathbf{V}}$ is the step size and is given by

$$\tau_{\mathbf{V}} = \begin{cases} \frac{\|\mathbf{V}-\mathbf{V}'\|_F^2}{\langle \mathbf{V}-\mathbf{V}', \nabla_{\mathbf{V}}\mathcal{L}(\mathbf{V})-\nabla_{\mathbf{V}}\mathcal{L}(\mathbf{V}') \rangle}, & \text{in odd-numbered iterations,} \\ \frac{\langle \mathbf{V}-\mathbf{V}', \nabla_{\mathbf{U}}\mathcal{L}(\mathbf{V})-\nabla_{\mathbf{V}}\mathcal{L}(\mathbf{V}') \rangle}{\|\nabla_{\mathbf{V}}\mathcal{L}(\mathbf{V})-\nabla_{\mathbf{V}}\mathcal{L}(\mathbf{V}')\|_F^2}, & \text{in even-numbered iterations.} \end{cases} \tag{8.23}$$

Here, $\mathbf{V}'$ is the counterpart of $\mathbf{V}$ obtained at the previous Stiefel manifold gradient descent iteration, $\tau_{\mathbf{V}} \in [\tau_{\min}, \tau_{\max}]$ with $\tau_{\min}$ and $\tau_{\max}$ being the minimum and maximum step sizes, respectively.

Given the convexity of (8.13), the Stiefel manifold gradient descent can certainly converge. $\mathbf{U}^l = [\mathbf{u}_1, \mathbf{V}^l]$, the solution of $\mathbf{U}$ at the l-th iteration of the DCA, can be obtained. $\mathbf{U}^l$ is taken as the local point in the following $(l+1)$-th iteration of DCA. This process repeats until the convergence of the DCA, and $\mathbf{U}^{(i)}$ is obtained.

8.3.2.2   Obfuscate Latent Spectral-Domain Stimulus S

Given $\mathbf{U}^{(i)}$, problem (8.9) is reformulated to an unconstrained DC program:

$$\min_{\mathbf{S}} \; g(\mathbf{S}) - h(\mathbf{S}), \tag{8.24}$$

where $g(\mathbf{S}) = \alpha^{(i-1)} \left\| \mathbf{U}^{(i)}\mathbf{S} - \mathbf{Y} \right\|_F^2$ and $h(\mathbf{S}) = \beta \left\| \mathbf{S} - \mathbf{S}^* \right\|_F^2$, both of which are convex functions. As a result, (8.24) can be solved by the DCA. More specifically, at the l-th iteration of the DCA ($l = 1, 2, \cdots$), the optimization problem below is addressed:

$$\min_{\mathbf{S}} \; \mathcal{F}(\mathbf{S}) = g(\mathbf{S}) - \langle \mathbf{S}, \nabla h(\mathbf{S}^{l-1}) \rangle, \tag{8.25}$$

where $\nabla h(\mathbf{S}^{l-1})$ is a gradient of $h(\mathbf{S})$ at $\mathbf{S}^{l-1}$, the local point obtained in the $(l-1)$-th iteration of the DCA. Given its convexity, problem (8.25) is solved using the gradient descent method. Specifically, $\mathbf{S}$ is updated by

$$\mathbf{S} = \mathbf{S}' - \tau_{\mathbf{S}} \nabla \mathcal{F}(\mathbf{S}'), \tag{8.26}$$

where $\mathbf{S}'$ represents the matrix of $\mathbf{S}$ at the previous iteration of the gradient descent, $\nabla \mathcal{F}(\mathbf{S})$ is the gradient of $\mathcal{F}(\mathbf{S})$, and $\tau_{\mathbf{S}}$ is the step size. Upon the convergence of the gradient descent method, $\mathbf{S}^l$ is obtained and used as the local point of the following $(l+1)$-th iteration of the DCA. Upon the convergence of the DCA, $\mathbf{S}^{(i)}$ is obtained.

8.3.2.3   Update α

Given $\mathbf{U}^{(i)}$ and $\mathbf{S}^{(i)}$, the value of α can be updated for the next AO iteration, i.e., $\alpha^{(i+1)}$, as given by

$$\alpha^{(i+1)} = \frac{\left\| \mathbf{U}^{(i)} - \mathbf{U}^* \right\|_F^2 + \beta \left\| \mathbf{S}^{(i)} - \mathbf{S}^* \right\|_F^2}{\left\| \mathbf{U}^{(i)}\mathbf{S}^{(i)} - \mathbf{Y} \right\|_F^2}. \tag{8.27}$$

8.3.3   Complexity Analysis

Algorithm 2 summarizes the proposed AO-based algorithm that solves problem (8.9), the above three parts are conducted in an alternating manner until the convergence of α, i.e., $\|\alpha^{(i+1)} - \alpha^{(i)}\| \leq \xi_{\mathrm{AO}}$ with $\xi_{\mathrm{AO}} \to 0$ being a preconfigured convergence accuracy. Each AO iteration starts by running the DCA and the Stiefel manifold gradient descent for $\mathbf{U}$ till convergence, followed by the DCA and the gradient descent for $\mathbf{S}$. Their respective convergence criteria are

$$\|\nabla_{\mathbf{V}}\mathcal{L}(\mathbf{V})\|_F \leq \xi_{\mathrm{SM}}, \; \|\nabla \mathcal{F}(\mathbf{S})\|_F \leq \xi_{\mathrm{GD}}, \tag{8.28}$$

$$\|\mathbf{U}^{l+1} - \mathbf{U}^l\| \leq \xi_{\mathrm{DC}}, \text{ and } \|\mathbf{S}^{l+1} - \mathbf{S}^l\| \leq \xi_{\mathrm{DC}}, \tag{8.29}$$

where ξ_{SM}, ξ_{GD}, and ξ_{DC} are the preconfigured accuracies for the Stiefel manifold gradient descent, gradient descent, and DCA, respectively.

Algorithm 2 The proposed AO-based obfuscation method for graph-structured data

---

1: **Initialization:** Set ϵ_{AO}, ξ_{SM}, ξ_{GD}, ξ_{DC}, μ, and $l = 0$; randomly initialize $\mathbf{U}^0$ and $\mathbf{S}^0$; calculate α^0 by (8.27);
2: **While** α is yet to converge with accuracy of ξ_{AO} **do**
3:     **for** $l = 0, 1, 2, \cdots$ **do**
4:         **While** $\|\nabla_{\mathbf{V}}\mathcal{L}(\mathbf{V})\|_F \geq \xi_{SM}$ **do**
5:             Update $\tau_{\mathbf{V}}$ by (8.23);
6:             Compute $\tau = \max(\min(\tau_{\mathbf{V}}, \tau_{\max}), \tau_{\min})$;
7:             Update $\mathbf{V}$ by (8.21);
8:             $\mathbf{V}' \leftarrow \mathbf{V}$;
9:         **end**
10:         **if** $\mathbf{U}^l = \mathbf{U}^{l-1}$ **then**
11:             Return to $\mathbf{U}^l$;
12:         **else**
13:             $\mathbf{U}^{l-1} \leftarrow \mathbf{U}^l$, $l = l + 1$;
14:         **end**
15:     **end**

16:     **for** $l = 0, 1, 2, \cdots$ **do**
17:         **While** $\|\nabla\mathcal{F}(\mathbf{S})\|_F \geq \xi_{GD}$ **do**
18:             Update $\mathbf{S}$ by (8.26);
19:             $\mathbf{S}' \leftarrow \mathbf{S}$;
20:         **end**
21:         **if** $\mathbf{S}^l = \mathbf{S}^{l-1}$ **then**
22:             Return to $\mathbf{S}^l$;
23:         **else**
24:             $\mathbf{S}^{l-1} \leftarrow \mathbf{S}^l$, $l = l + 1$;
25:         **end**
26:     **end**
27:     Update α by (8.27);
28: **end.**

In each iteration of Algorithm 2, the cost of evaluating $\mathbf{U}$ is primarily influenced by the SVD in (8.21) per Stiefel manifold gradient descent iteration, incurring a cost of $\mathcal{O}(N^3)$. The complexity of evaluating $\mathbf{S}$ is $\mathcal{O}(MN)$ per gradient descent iteration, dominated by calculating the gradient of $\mathcal{F}(\mathbf{S})$. Therefore, the overall cost of Algorithm 2 is $\mathcal{O}\Big((MN\log(\frac{1}{\xi_{GD}}) + N^3\log(\frac{1}{\xi_{SM}}))\log(\frac{1}{\xi_{DC}})\log(\frac{1}{\xi_{AO}})\Big)$, where $\log(\frac{1}{\xi_{SM}})$, $\log(\frac{1}{\xi_{GD}})$, $\log(\frac{1}{\xi_{DC}})$, and $\log(\frac{1}{\xi_{AO}})$ give the numbers of iterations for the Stiefel manifold gradient, gradient descent, DCA, and AO to converge, respectively.

8.4   Numerical Evaluation

Extensive simulations and experiments are carried out to gauge the proposed obfuscation technique for graph data, i.e., Algorithm 1, using a desktop computer with an i7-8650U CPU and 16G RAM. By default, it sets $\xi_{AO} = \xi_{SM} = \xi_{GD} = \epsilon_{DC} = 10^{-5}$.

Graph Inference Attack: The resistance of the new graph data obfuscation algorithm to graph interference attacks is assessed. Following is a list of the latest graph inference techniques that can be used to launch graph interference attacks.

- Dong's algorithm[44]: This is an alternating minimization algorithm that fixes one variable and solves the others in an alternating manner to infer the optimal Laplacian matrix under the assumption of smooth data.

- Kalofolias' algorithm [86]: This primal-dual algorithm was designed to learn the weighted adjacency matrix of graphs.

- Sardellitti's TV graph learning algorithm [140]; This is a two-step strategy consisting of (a) deriving the GFT basis out of data via AO and (b) then learning the graph Laplacian matrix using the sparsifying transform via convex optimization technique, named "TV-based graph learning."

- Sardellitti's ESA graph learning algorithm [140]: Different from the TV-based graph learning algorithm, this algorithm introduces a two-step strategy by deriving the graph Laplacian using sparsifying transform, as well as the GFT basis utilizing convex optimization, named "ESA graph learning" in the second step.

- Humbert's algorithm [80]: This method runs Riemannian manifold gradient descent and linear cone programs in an alternating fashion. It learns graphs from multivariate data with smoothness and band-limitedness.

Performance Metric: F-measure is used to measure the graph learning results and is denoted as

$$\text{F-measure} = 2 \cdot \frac{\text{Precision} \cdot \text{Recall}}{\text{Precision} + \text{Recall}}, \tag{8.30}$$

where Precision $= \mathcal{E}_g \cap \mathcal{E}_r / \mathcal{E}_r$ and Recall $= \mathcal{E}_g \cap \mathcal{E}_r / \mathcal{E}_g$. Here, $\mathcal{E}_g$ and $\mathcal{E}_r$ collect ground-truth and detected graphs, respectively.

The correlation coefficient $\rho_{\mathbf{W}}(\mathbf{W}_0, \mathbf{W})$ (or $\rho_{\mathbf{W}}$) between a graph detected by the graph learning attacks and its corresponding ground-truth is expressed as [140]

$$\rho_{\mathbf{W}}(\mathbf{W}_0, \mathbf{W}) = \frac{\sum_{ij} W_{0ij} W_{ij}}{\sqrt{\sum_{ij} W_{0ij}^2} \sqrt{\sum_{ij} W_{ij}^2}}, \tag{8.31}$$

where $\mathbf{W}$ is the weighted adjacency matrix of a detected graph with the (i, j)-th element W_{ij}, and $\mathbf{W}_0$ is the weighted adjacency matrix of the corresponding ground-truth graph with the (i, j)-th element W_{0ij}.

DP-based benchmark: The DP-based Gaussian mechanism is considered as the benchmark for the proposed Algorithm 2. In the DP-based mechanism, the DP noise yields the Gaussian distribution with the variance of the noise given by [180]

$$\sigma_{\mathrm{DP}}^2 = \frac{2 \ln (1.25/\delta) \, \Delta^2}{\epsilon^2}, \tag{8.32}$$

where δ is the probability of information accidentally being leaked, ϵ is the privacy budget, and Δ is the global sensitivity.

As discussed in Section 8.1, no existing technique has been designed to protect the latent information of graph-structured data in the existing literature. While DP is a general approach to obfuscating data with a balance between privacy and utility, it is unsuitable for graph-structured data, e.g., brain network data. Specifically, adding DP noises to the GFT basis, $\mathbf{U}^*$, is unacceptable and would breach the constraints, e.g., the orthonormality of the GFT basis.

DP is applied to the observed graph-structured data $\mathbf{Y}$. Here, $\Delta = \frac{\|\mathbf{U}-\mathbf{U}^*\|_F^2 + \beta \|\mathbf{S}-\mathbf{S}^*\|_F^2}{\|\mathbf{US}-\mathbf{Y}\|_F^2}$. However, the direct use of DP to perturb the graph-structured data can be ineffective in protecting the latent information, because the GFT basis characterizing the latent graph structure is stringently constrained and less susceptible to the added noises, as discussed below.

Remark 7 *The latent graph structures of graph data are resistant to the Gaussian DP noise. This is due to the fact that the GFT basis $\mathbf{U}^*$ is solely determined by the eigenvectors of $\mathbf{YY}^T$, while $\hat{\mathbf{Y}}\hat{\mathbf{Y}}^T - M\sigma_{\mathrm{DP}}^2\mathbf{I}$ is an asymptotic unbiased estimator of $\mathbf{YY}^T$ and $\hat{\mathbf{Y}}\hat{\mathbf{Y}}^T$ has the same eigenvectors as $\hat{\mathbf{Y}}\hat{\mathbf{Y}}^T - M\sigma_{\mathrm{DP}}^2\mathbf{I}$. Here, $\hat{\mathbf{Y}} = \mathbf{Y} + \mathbf{n}$ is the perturbed version of $\mathbf{Y}$ with the DP noise added. $\mathbf{n} \sim \mathcal{N}(0, \sigma_{\mathrm{DP}}^2\mathbf{I})$ is the Gaussian DP noise. Specifically, the expectation of $\hat{\mathbf{Y}}\hat{\mathbf{Y}}^T$ over the Gaussian DP noise is given by*

$$\mathbb{E}(\hat{\mathbf{Y}}\hat{\mathbf{Y}}^T) = \mathbf{YY}^T + M\sigma_{\mathrm{DP}}^2\mathbf{I}. \tag{8.33}$$

Hence, the eigenvectors of $\mathbb{E}(\hat{\mathbf{Y}}\hat{\mathbf{Y}}^T)$ and $\mathbf{YY}^T$ are identical.

Without the DP noise, $\mathbf{U}^*$ in (8.6) is solely determined by the eigenvectors of $\mathbf{YY}^T$. In the presence of the DP noise, $\mathbf{U}^*$ is given by

$$\mathbf{U}^* = \mathrm{Eig} \left[(\mathbf{I} - \mathbf{u}_1\mathbf{u}_1^T) \, \mathbb{E}(\hat{\mathbf{Y}}\hat{\mathbf{Y}}^T) \left(\mathbf{I} - \mathbf{u}_1\mathbf{u}_1^T \right)^T \right]. \tag{8.34}$$

Since $\mathbb{E}(\frac{1}{M}\hat{\mathbf{Y}}\hat{\mathbf{Y}}^T - \sigma_{\mathrm{DP}}^2\mathbf{I}) = \frac{1}{M}\mathbf{YY}^T$ (i.e., $\frac{1}{M}\hat{\mathbf{Y}}\hat{\mathbf{Y}}^T - \sigma_{\mathrm{DP}}^2\mathbf{I}$ is an asymptotic unbiased estimate of $\frac{1}{M}\mathbf{YY}^T$ as $M \to \infty$) based on (8.33) and the eigenvectors of $\hat{\mathbf{Y}}\hat{\mathbf{Y}}^T - M\sigma_{\mathrm{DP}}^2\mathbf{I}$ and $\hat{\mathbf{Y}}\hat{\mathbf{Y}}^T$ are identical, $\mathbf{U}^*$ is asymptotically approximately given by

$$\mathbf{U}^* \approx \mathrm{Eig} \left[(\mathbf{I} - \mathbf{u}_1\mathbf{u}_1^T) \, \hat{\mathbf{Y}}\hat{\mathbf{Y}}^T \left(\mathbf{I} - \mathbf{u}_1\mathbf{u}_1^T \right)^T \right]. \tag{8.35}$$

In this sense, perturbing the graph-structured data $\mathbf{Y}$ *is less effective in perturbing the latent graph structures of* $\mathbf{Y}$.

8.4.1 Performance Evaluation with Synthetic Data

Random graphs with six connections per node are generated by using the widely adopted Random Geometric model [37].

- *Ground truth:* With a graph created using the Random Geometric graph model, its ground-truth Laplacian is derived, denoted by $\mathbf{L}_0$. The ground-truth GFT basis is also obtained, denoted by $\mathbf{U}_0$, by taking the SVD of $\mathbf{L}_0$.

- *Synthetic data:* The observed graph data is produced $\mathbf{Y} = \mathbf{U}_0 \mathbf{S}_0$ with $\mathbf{S}_0 = [\mathbf{s}_{0,1}, \cdots, \mathbf{s}_{0,M}] \in \mathbb{R}^{N \times M}$ randomly generated yielding $\mathbf{s}_{0,m} \sim \mathcal{N}(0, \mathbf{\Lambda}_K^\dagger)$, where $\mathrm{diag}(\mathbf{\Lambda}_K) = (\lambda_1, \cdots, \lambda_K, 0, \cdots, 0)$. The precision matrix of $\mathbf{s}_{0,m}$ is defined as the eigenvalue matrix of $\mathbf{L}$ with the most significant $(N - K)$ eigenvalues replaced by 0, as done in [91].

- Obfuscated data (proposed Algorithm 2): The observable graph data $\mathbf{Y}$ is obfuscated using Algorithm 2, where $\mathbf{Y}$ is the input to the algorithm first to detect and then obfuscate $\mathbf{U}^*$ and $\mathbf{S}^*$. The obfuscated versions of $\mathbf{U}^*$ and $\mathbf{S}^*$ are $\mathbf{U}$ and $\mathbf{S}$, respectively.

- Obfuscated data (DP-based benchmark): For a fair comparison with Algorithm 2, the variance of the convergent value of $||\mathbf{US} - \mathbf{Y}||_F$ is gauged under Algorithm 2. Then, $\mathbf{Y}$ is obfuscated by adding the DP noise to $\mathbf{Y}$, i.e., $\mathbf{Y} + \mathbf{n}_{\mathrm{DP}}$, where $\mathbf{n}_{\mathrm{DP}} \sim \mathcal{N}(0, \sigma^2 \mathbf{I})$. In this way, the utility of $\mathbf{Y}$ can be consistent between the proposed Algorithm 2 and the DP-based benchmark.

1) *Convergence Analysis:* Fig. 8.2 plots the convergence behaviors of Algorithm 2 with an increasing number of iterations, where $K = 15$, $M = 300$, and $\beta = 0.1, 0.5, 1.0, 1.5$, and 2.0. Fig. 8.2(a) plots the auxiliary variable, α, under different values of β. Figs. 8.2(b)–8.2(d) plot the changes of $||\mathbf{U} - \mathbf{U}^*||_F$, $||\mathbf{S} - \mathbf{S}^*||_F$, and $||\mathbf{US} - \mathbf{Y}||_F$, respectively. It is observed in Figs. 8.2(a) and 8.2(c) that α and $||\mathbf{S} - \mathbf{S}^*||_F$ grow quickly at the beginning and then gradually converge at around the 20th and 50th iterations, respectively. It is also observed that $||\mathbf{U} - \mathbf{U}^*||_F$ first increases rapidly and then drops slowly and converges. By contrast, $||\mathbf{US} - \mathbf{Y}||_F$ first drops considerably, and then increases slightly and converges, as shown in Figs. 8.2(b) and 8.2(d).

It is also shown in Fig. 8.2 that α, $||\mathbf{S} - \mathbf{S}^*||_F$, $||\mathbf{U} - \mathbf{U}^*||_F$, and $||\mathbf{US} - \mathbf{Y}||_F$ change differently with the value of β. Specifically, α increases as β grows from 0.1 to 2.0. $||\mathbf{S} - \mathbf{S}^*||_F$ increases as β grows from 0.1 to 1.5 and decreases slightly when $\beta = 2.0$. However, $||\mathbf{U} - \mathbf{U}^*||_F$ and $||\mathbf{US} - \mathbf{Y}||_F$ show less clear dependence on β. Nevertheless, a reasonable value is chosen for β, i.e., $\beta = 1$, that can achieve the smallest value of $||\mathbf{US} - \mathbf{Y}||_F$ and meanwhile maintain large $||\mathbf{U} - \mathbf{U}^*||_F$ and $||\mathbf{S} - \mathbf{S}^*||_F$. Typically, β is specified empirically, depending on the relative importance or preference of the graph structure and stimulus. No additional experiments or computational overhead is needed for specifying β.

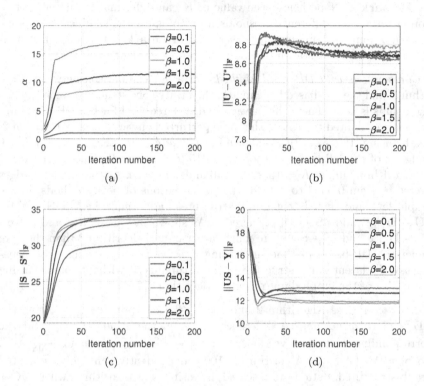

FIGURE 8.2: The convergence performance of Algorithm 2 under different values of β, where $K = 15$ and $M = 300$.

Fig. 8.3 plots the changes of $\|\mathbf{U} - \mathbf{U}^*\|_F$, $\|\mathbf{S} - \mathbf{S}^*\|_F$, and $\|\mathbf{US} - \mathbf{Y}\|_F$ with an increasing number of iterations under different values of K, where $N = 30$, and $\beta = 1$. It is observed that $\|\mathbf{U} - \mathbf{U}^*\|_F$, $\|\mathbf{S} - \mathbf{S}^*\|_F$, and $\|\mathbf{US} - \mathbf{Y}\|_F$ increase with K. In other words, the bigger K can lead to better obfuscations of $\mathbf{U}$ and $\mathbf{S}$, but the utility of $\mathbf{Y}$ can be penalized as a cost. It is also observed that the growths rates of $\|\mathbf{U} - \mathbf{U}^*\|_F$, $\|\mathbf{S} - \mathbf{S}^*\|_F$, and $\|\mathbf{US} - \mathbf{Y}\|_F$ decrease with the increase of K. This is because when K is large enough, the graph learned from the observed data contains almost all information, and a further increase of K has little impact on $\|\mathbf{U} - \mathbf{U}^*\|_F$, $\|\mathbf{S} - \mathbf{S}^*\|_F$, and $\|\mathbf{US} - \mathbf{Y}\|_F$.

It is noted that K is the bandwidth of the observed graph-structured data $\mathbf{Y}$, and is not a parameter of the proposed obfuscation algorithm, i.e., Algorithm 2. The correct identification of K is important to correctly extract the latent information, especially in the presence of non-negligible observation noises. The value of K can be experimentally specified at additional computational overhead, when extracting the latent information $\mathbf{U}^*$ and $\mathbf{S}^*$ from $\mathbf{Y}$. The additional overhead is negligible due to the closed-form expressions derived, i.e., (8.6) and (8.7), that can be used directly to specify K; see **Remark 1**. Nonetheless, the value of K can differ under different detection criteria, e.g., different thresholds used to assess the significance of each spectrum-domain component, as described in **Remark 1**. For generality, multiple values for K are taken to assess the proposed algorithm.

2) *Comparisons with DP-based Method:* For a fair comparison between Algorithm 2 and the DP-based benchmark, the utility consistency between the two algorithms is kept. Then, the privacy budget ϵ required for the DP-based approach is evaluated to achieve the same perturbation variance as Algorithm 2, i.e., by using [180, Eq. (2)]. As shown in Fig. 8.4(c), the corresponding privacy budget ϵ of the DP-based method is 0.5315 to achieve the same perturbation as Algorithm 2 under $K = 15$, or 0.4946 under $K = 20$. As the privacy budget decreases from 0.5315 to 0.4946, the perturbations of both methods on the graph data and their latent graph structures and stimuli, i.e., $\|\mathbf{US} - \mathbf{Y}\|_F$, $\|\mathbf{U} - \mathbf{U}^*\|_F$, and $\|\mathbf{S} - \mathbf{S}^*\|_F$, increase. As shown in Figs. 8.4(a) and 8.4(b), $\|\mathbf{U} - \mathbf{U}^*\|_F$ and $\|\mathbf{S} - \mathbf{S}^*\|_F$ are much larger under Algorithm 2 than they are under the DP-based method, indicating the significantly stronger perturbations of the latent graph structure $\mathbf{U}^*$ and stimulus $\mathbf{S}^*$ without compromising the utility of the graph data $\mathbf{Y}$ under Algorithm 2.

Fig. 8.5 compares Algorithm 2 and the DP-based approach by plotting the CDFs of their normalized differences between the ground truths and the corresponding perturbed versions, i.e., $\frac{\|\mathbf{U}-\mathbf{U}^*\|_F}{\|\mathbf{U}^*\|_F}$, $\frac{\|\mathbf{S}-\mathbf{S}^*\|_F}{\|\mathbf{S}^*\|_F}$, and $\frac{\|\mathbf{US}-\mathbf{Y}\|_F}{\|\mathbf{Y}\|_F}$. To plot the CDFs of Algorithm 2, 100 independently randomly generated graph-structured data is obfuscated by using the algorithm, where K is empirically specified and preconfigured. Under a given K, for each random graph-structured data, the variance of the perturbations on the graph data, i.e., $\|\mathbf{US} - \mathbf{Y}\|_F$, is evaluated.

To effectively compare Algorithm 2 with the latest DP-based method, we evaluate the privacy budget ϵ required for the DP-based method to achieve the same perturbation variance as Algorithm 2, i.e., by using [180, eq. 2]. Then, we perturb the random graph-structured data using the DP-based approach under the privacy budget and extract the latent information (i.e., graph structure and stimulus) from the DP-perturbed graph-structured data using the five state-of-the-art graph learning techniques with the K value (if needed). As a result of the randomness of the DP noise, the privacy budget ϵ is within $[0.3523, 0.8673]$ when $K = 15$, and within $[0.3269, 0.6857]$ when $K = 20$. It is observed that $\frac{\|\mathbf{U}-\mathbf{U}^*\|_F}{\|\mathbf{U}^*\|_F}$ and $\frac{\|\mathbf{S}-\mathbf{S}^*\|_F}{\|\mathbf{S}^*\|_F}$ are consistently better under Algorithm 2 than under the DP-based method, indicating the better obfuscation effect under Algorithm 2 than under the DP-based benchmark.

Fig. 8.6 plots $\frac{\|\mathbf{U}-\mathbf{U}^*\|_F}{\|\mathbf{U}^*\|_F}$ versus $\frac{\|\mathbf{S}-\mathbf{S}^*\|_F}{\|\mathbf{S}^*\|_F}$ under Algorithm 2 and the DP-based method, where $\mathbf{n}_{\mathrm{DP}}$ is consistent with Fig. 8.4(c) and the DP-based method is

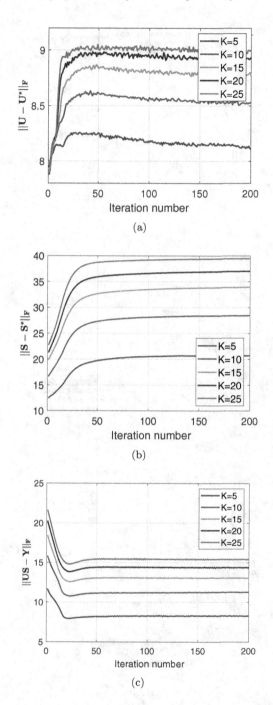

FIGURE 8.3: The performance of Algorithm1 under different values of K, where $\beta = 1.0$.

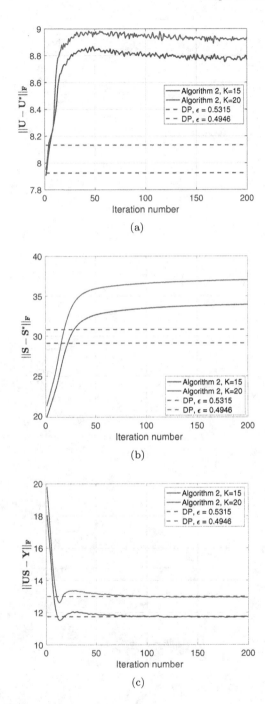

FIGURE 8.4: Comparison between Algorithm 2 and the DP-based method, where $\beta = 1$ and $K = 15, 20$.

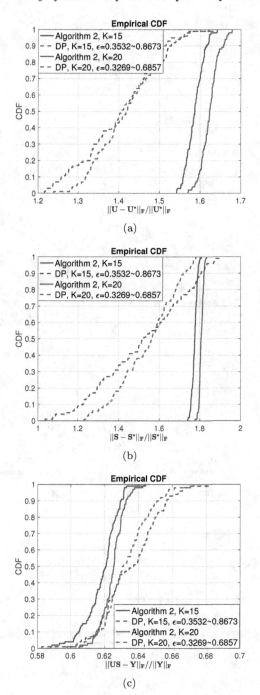

FIGURE 8.5: The CDFs of $\frac{\|\mathbf{U}-\mathbf{U}^*\|_F}{\|\mathbf{U}^*\|_F}$, $\frac{\|\mathbf{S}-\mathbf{S}^*\|_F}{\|\mathbf{S}^*\|_F}$, and $\frac{\|\mathbf{US}-\mathbf{Y}\|_F}{\|\mathbf{Y}\|_F}$ under Algorithm 2 and the DP-based benchmark, where $\beta = 1$ and $K = 15, 20$.

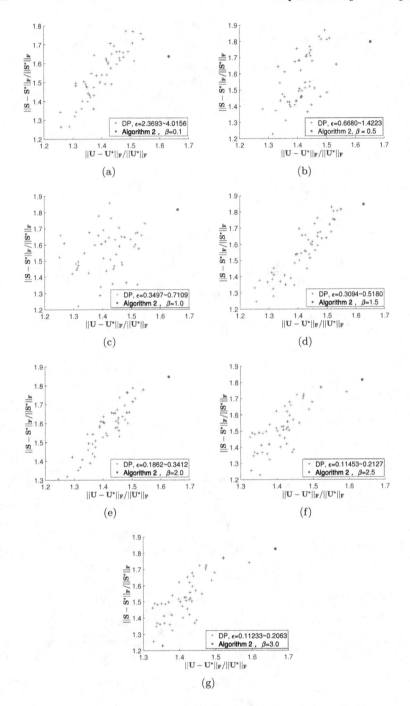

FIGURE 8.6: $\frac{\|\mathbf{U}-\mathbf{U}^*\|_F}{\|\mathbf{U}^*\|_F}$ versus $\frac{\|\mathbf{S}-\mathbf{S}^*\|_F}{\|\mathbf{S}^*\|_F}$, where Algorithm 2 and the DP-based method are compared under different β values and $K=20$.

repeated independently for 50 times. It is observed that the DP-based benchmark gives less reliable obfuscation results than Algorithm 2, as it undergoes the randomness of the DP noise. For example, the normalized obfuscation level of $\mathbf{U}$ and $\mathbf{S}$ ranges from 1.3 to 1.8 under the DP-based benchmark, and it is around 1.7 under Algorithm 2, when the benchmark is expected to achieve consistent utility with Algorithm 2. It is also noticed that under Algorithm 1, $\frac{\|\mathbf{S}-\mathbf{S}^*\|_F}{\|\mathbf{S}^*\|_F}$ is small when $\beta = 0.1$. With the growth of β, $\frac{\|\mathbf{S}-\mathbf{S}^*\|_F}{\|\mathbf{S}^*\|_F}$ increases and tends to stabilize at around 1.85.

Algorithm 2 is further compared with the DP-based benchmark under different state-of-the-art graph learning techniques utilized for graph inference attacks, including Dong's algorithm [44], Kalofolias' algorithm [86], Sardellitti's TV algorithm [140], Sardellitti's ESA algorithm [140], and Humbert's algorithm [80].

For a fair comparison, all regularization parameters of the graph learning techniques are tested and optimized accordingly. Again, we keep consistent utilities between Algorithm 2 and the DP-based approach and evaluate their resistance to the latest graph inference attacks. Without loss of generality, $\beta = 1$ is set for Algorithm 2.

As shown in Table 8.1, Algorithm 2 is lower by 31.16% and 30.18% than the DP-based approach in F-measure and correlation coefficient $\rho_{\mathbf{W}}$, respectively. In other words, the graphs extracted from the perturbed graph-structured data using the latest graph learning techniques are significantly more dissimilar to their ground truths under Algorithm 2. As a result, the attack success rates of the graph inference attacks can be reduced dramatically.

Fig. 8.7 plots the CDFs of the F-measure and $\rho_{\mathbf{W}}$ values of Algorithm 2 and the DP-based approach under the five graph inference attacks. As shown in the figure, Algorithm 1 is substantially more resistant to the graph inference attacks than the DP-based method due to its significantly weaker similarity between the latent graph structures learned by the graph inference attacks and the ground truths.

Fig. 8.8 plots the attack success rates of the five graph inference attacks on Algorithm 2 and the DP-based approach, where the x-axis provides the threshold of F-measure or $\rho_{\mathbf{W}}$, above which a graph inference attack is treated as successful. As shown in the figure, Algorithm 2 allows dramatically lower attack success rates for the five attacks. Suppose that a graph inference attack is successful if the F-measure is larger than 0.5; i.e., the F-measure threshold is 0.5. The attack success rates of all five graph interference attacks are 100% under the DP-based approach. By contrast, Algorithm 2 resists all five attacks with an attack success rate of zero.

Table 8.1: Comparison in utility ($\|\mathbf{US} - \mathbf{Y}\|_F$) and resistance (F-measure and $\rho\mathrm{w}$) between Algorithm 2 and the DP-based benchmark under graph inference attacks launched by the considered five graph learning techniques. $K = 15$, $N = 30$, $M = 300$, and $\beta = 1$.

	Dong [35]	Kalofolias [36]	Sardellitti-TV [9]	Sardellitti-ESA [9]	Humbert [10]
Ground-truth					
F-measure	0.8517 ($\pm$ 0.025)	0.8642 ($\pm$ 0.022)	0.8125 ($\pm$ 0.031)	0.8571 ($\pm$ 0.040)	0.9053 ($\pm$ 0.019)
$\rho\mathrm{w}$	0.9037 ($\pm$ 0.017)	0.9102 ($\pm$ 0.019)	0.9121 ($\pm$ 0.020)	0.9274 ($\pm$ 0.027)	0.9217 ($\pm$ 0.035)
DP					
F-measure	0.7989 ($\pm$ 0.021)	0.7671 ($\pm$ 0.039)	0.7927 ($\pm$ 0.037)	0.7604 ($\pm$ 0.026)	0.7737 ($\pm$ 0.043)
$\rho\mathrm{w}$	0.8331 ($\pm$ 0.037)	0.8035 ($\pm$ 0.034)	0.8288 ($\pm$ 0.029)	0.8039 ($\pm$ 0.046)	0.7865 ($\pm$ 0.037)
$\|\mathbf{U} - \mathbf{U}^*\|_F$	7.8133 ($\pm$ 0.119)	7.9537 ($\pm$ 0.127)	7.8047 ($\pm$ 0.106)	7.7952 ($\pm$ 0.201)	7.9249 ($\pm$ 0.131)
$\|\mathbf{S} - \mathbf{S}^*\|_F$	29.113 ($\pm$ 0.157)	28.201 ($\pm$ 0.179)	29.126 ($\pm$ 0.211)	28.213 ($\pm$ 0.144)	28.755 ($\pm$ 0.207)
$\|\mathbf{US} - \mathbf{Y}\|_F$	11.925 ($\pm$ 0.361)	11.593 ($\pm$ 0.296)	12.449 ($\pm$ 0.407)	12.622 ($\pm$ 0.418)	12.364 ($\pm$ 0.379)
Algorithm 2					
F-measure	**0.4873** ($\pm$ **0.011**)	**0.4057** ($\pm$ **0.026**)	**0.3869** ($\pm$ **0.014**)	**0.3627** ($\pm$ **0.021**)	**0.4570** ($\pm$ **0.015**)
$\rho\mathrm{w}$	**0.5313** ($\pm$ **0.031**)	**0.4263** ($\pm$ **0.032**)	**0.4119** ($\pm$ **0.022**)	**0.3757** ($\pm$ **0.025**)	**0.4652** ($\pm$ **0.024**)
$\|\mathbf{U} - \mathbf{U}^*\|_F$	**8.7998** ($\pm$ **0.125**)	**8.8213** ($\pm$ **0.241**)	**8.8057** ($\pm$ **0.209**)	**8.8186** ($\pm$ **0.117**)	**8.8112** ($\pm$ **0.126**)
$\|\mathbf{S} - \mathbf{S}^*\|_F$	**33.831** ($\pm$ **0.241**)	**34.223** ($\pm$ **0.109**)	**33.982** ($\pm$ **0.212**)	**34.351** ($\pm$ **0.153**)	**34.174** ($\pm$ **0.113**)
$\|\mathbf{US} - \mathbf{Y}\|_F$	**11.743** ($\pm$ **0.195**)	**11.307** ($\pm$ **0.228**)	**12.013** ($\pm$ **0.231**)	**12.147** ($\pm$ **0.212**)	**11.599** ($\pm$ **0.239**)

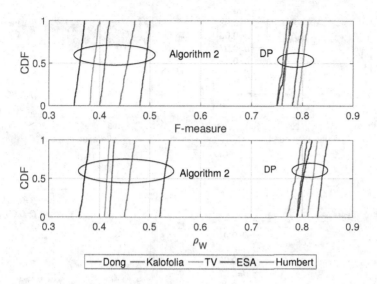

FIGURE 8.7: CDFs of Algorithm 2 and the DP-based method, where $K = 15$, $N = 30$, $M = 300$, and $\beta = 1$.

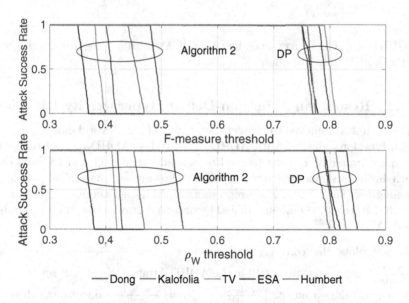

FIGURE 8.8: Attack success rates of Algorithm 2 and the DP-based method, where $K = 15$, $N = 30$, $M = 300$, and $\beta = 1$.

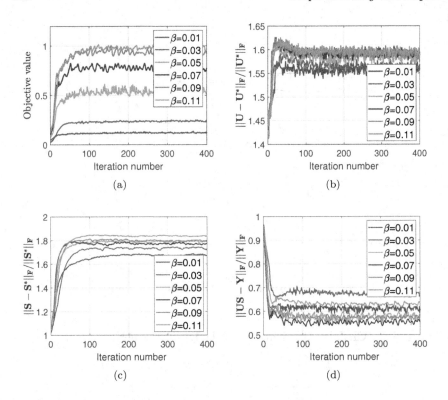

FIGURE 8.9: The convergence behavior of Algorithm 2 regarding subject 10 of the ADHD dataset, where $K = 55$.

8.4.2 Results on Attention-Deficit Hyperactivity Disorder

Algorithm 2 is employed to protect the privacy of a real-world dataset, i.e., the brain functional dataset with ADHD. The studied ADHD dataset comprises the data sample of 42 right-handed boys aged between 11 and 16 years old[1]. Each brain is segmented into 90 ROIs, following an anatomical automatic labeling template. Every node represents an ROI, and there are 232 samples per ROI in the time domain [155]. The observed graph-structured brain data yields $\mathbf{Y} \in \mathbb{R}^{90 \times 232}$.

Fig. 8.9 plots the convergence behaviors of α, $\frac{||\mathbf{U}-\mathbf{U}^*||_F}{||\mathbf{U}^*||_F}$, $\frac{||\mathbf{S}-\mathbf{S}^*||_F}{||\mathbf{S}^*||_F}$, and $\frac{||\mathbf{US}-\mathbf{Y}||_F}{||\mathbf{Y}||_F}$ regarding subject 10 in the ADHD dataset, where K is set as 55 and different β values are tested. $\frac{||\mathbf{U}-\mathbf{U}^*||_F}{||\mathbf{U}^*||_F}$ versus $\frac{||\mathbf{S}-\mathbf{S}^*||_F}{||\mathbf{S}^*||_F}$ concerning subject 10 of the dataset is also plotted in Fig. 8.10. Consistent observations are made

---

[1]The dataset is obtained from the ADHD-200 global competition database (https://www.nitrc.org/projects/neurobureau/).

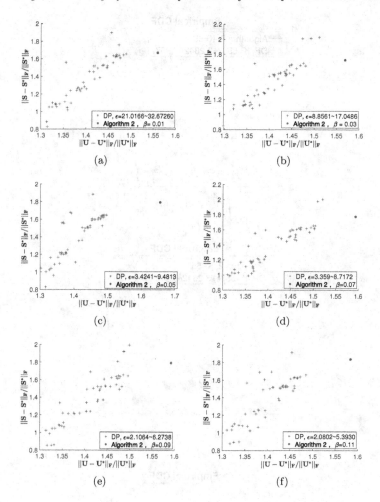

FIGURE 8.10: $\frac{\|\mathbf{U}-\mathbf{U}^*\|_F}{\|\mathbf{U}^*\|_F}$ versus $\frac{\|\mathbf{S}-\mathbf{S}^*\|_F}{\|\mathbf{S}^*\|_F}$ for subject 10, where Algorithm 2 and the DP-based method are compared under different values of β, and $K = 55$.

here with synthetic data (see Figs. 8.2 and 8.6), indicating the reliability of the Algorithm 2 in practical scenarios.

The CDF of $\frac{\|\mathbf{U}-\mathbf{U}^*\|_F}{\|\mathbf{U}^*\|_F}$, $\frac{\|\mathbf{S}-\mathbf{S}^*\|_F}{\|\mathbf{S}^*\|_F}$, and $\frac{\|\mathbf{US}-\mathbf{Y}\|_F}{\|\mathbf{Y}\|_F}$ by considering all subjects in the ADHD dataset are plotted in Fig. 8.11. It shows that the Algorithm 2 consistently achieves better performance than the DP-based method by providing more significant obfuscations on $\mathbf{U}$ and $\mathbf{S}$ in Figs. 8.11(a) and 8.11(b) while maintaining the consistent utility $\mathbf{Y}$ in Fig. 8.11(c).

Using the graph learning algorithm developed in [140], namely, Sardellitti-TV [140], the brain functional networks from the observed data $\mathbf{Y}$ can be

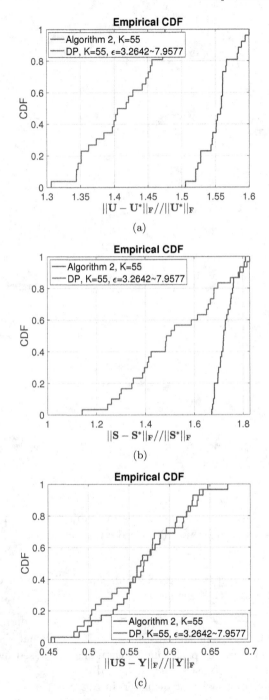

FIGURE 8.11: The CDFs of $\frac{||\mathbf{U}-\mathbf{U}^*||_F}{||\mathbf{U}^*||_F}$, $\frac{||\mathbf{S}-\mathbf{S}^*||_F}{||\mathbf{S}^*||_F}$, and $\frac{||\mathbf{US}-\mathbf{Y}||_F}{||\mathbf{Y}||_F}$ regarding all subjects in the ADHD dataset.

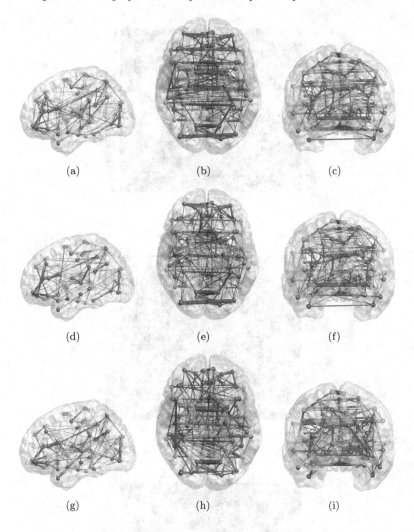

FIGURE 8.12: Visualization of the brain network from the sagittal, axial, and coronal views using BrainNet Viewer. The learned graphs from the data of (a)-(c) the original observation, (d)-(f) the DP-based method, (g)-(i) the proposed Algorithm 2.

learned. Then, the GFT basis $\mathbf{U}$ and the stimulus $\mathbf{S}$ are obfuscated by using Algorithm 2 and its DP-based benchmark. Both the original and the obfuscated brain networks, connecting the ROIs are constructed as weighted adjacent coefficients. The learned brain networks are visualized by the BrainNet Viewer toolbox [168].

Fig. 8.12 visualizes the learned brain functional networks (from the ground truths, and the obfuscated versions based on Algorithm 2 and the DP-based

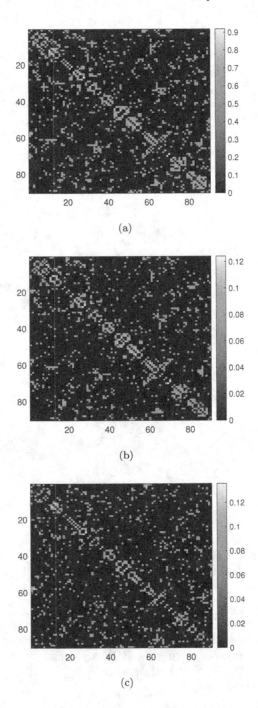

FIGURE 8.13: The weighted adjacency matrices of the learned graphs from the data of (a) the original observation, (b) the DP-based method, and (c) the proposed Algorithm 2.

Table 8.2: Comparison between Algorithm 2 and the DP-based method under graph inference attacks launched by the five considered graph learning algorithms on the ADHD dataset, where $K = 55$ and $\beta = 0.05$.

	Dong [44]	Kalofolias [86]	Sardellitti-TV [140]	Sardellitti-ESA [140]	Humbert [80]
DP					
F-measure	0.8187 ($\pm$ 0.021)	0.7832 ($\pm$ 0.047)	0.7523 ($\pm$ 0.045)	0.7523 ($\pm$ 0.028)	0.7773 ($\pm$ 0.033)
ρ_w	0.8431 ($\pm$ 0.037)	0.8035 ($\pm$ 0.029)	0.7769 ($\pm$ 0.045)	0.7617 ($\pm$ 0.039)	0.7846 ($\pm$ 0.053)
Algorithm 2					
F-measure	0.3232 ($\pm$ 0.042)	0.3956 ($\pm$ 0.018)	0.3884 ($\pm$ 0.042)	0.3956 ($\pm$ 0.034)	0.3972 ($\pm$ 0.027)
ρ_w	0.3111 ($\pm$ 0.027)	0.4137 ($\pm$ 0.034)	0.3441 ($\pm$ 0.037)	0.3962 ($\pm$ 0.039)	0.4018 ($\pm$ 0.032)

benchmark) from three different views of a brain, i.e., the sagittal view, axial view, and coronal view. To better display the differences between the brain functional networks learned from the ground truth and its obfuscated versions, the brain functional networks are transformed into the weighted adjacency matrices and shown in Fig. 8.13. It is found that the weighted adjacency matrix learned from the obfuscated version based on Algorithm 2 differs more substantially from that learned from the ground truth, compared to the obfuscated version based on the DP-based benchmark. In other words, Algorithm 2 can more effectively perturb the GFT basis and stimuli with little impact on the utility of the observed data.

Table 8.2 compares Algorithm 2 and the DP-based method in their resistance to the latest graph interference attacks on all subjects in the ADHD dataset. F-measure and $\rho_{\mathbf{W}}$ are used to measure the resistance. For each graph inference method considered, the learned graphs from the original observed data are taken as the ground truths, and then obfuscated using Algorithm 2 and the DP-based benchmark. As shown in the table, Algorithm 1 is more robust to all the considered graph inference attacks than the DP-based method, achieving lower mean scores in both F-measure and $\rho_{\mathbf{W}}$. In particular, Table 8.2 shows that Algorithm 2 is at least 36.39% and 36.55% better than the DP-based benchmark in F-measure and $\rho_{\mathbf{W}}$, respectively.

8.5   Concluding Remark

This chapter delivered a new privacy perturbation algorithm to protect the privacy of the latent graph structures and stimuli of graph-structured data. Specifically, a new multi-objective problem is formulated, and the limitation of DP is revealed. The new algorithm was developed to decouple the multi-objective problem and obfuscate the latent graph structures and stimuli in an alternating manner. Experiments performed on synthetic graph data and the practical ADHD dataset demonstrated that the presented approach can significantly outperform its DP-based benchmark under graph inference attacks. The approach can have extensive applications to personal health records, finance transactions, and other graph-structured data.

9

Future Directions and Challenges

The chapter discusses upcoming trends and challenges in the fields of privacy and graph learning. It highlights emerging trends in privacy and graph learning, including novel approaches and applications. The chapter also addresses the difficulties in scaling privacy-preserving techniques to handle large datasets and complex systems. Finally, it outlines potential future research directions to advance these fields, emphasizing areas where further innovation and exploration are needed.

9.1 Emerging Trends in Privacy and Graph Learning

Emerging trends in privacy and graph learning are significantly influencing how data privacy is managed and how relationships within data are analyzed. Novel approaches and applications are being developed to address the complex challenges in this evolving field.

One prominent trend is the integration of differential privacy with graph learning algorithms [121]. Differential privacy provides strong privacy guarantees by adding controlled noise to the data, making it difficult to identify individual data points. When applied to graph learning, differential privacy ensures that the privacy of individuals within the graph is protected even as the graph's structural and relational patterns are analyzed. Researchers are developing differential privacy techniques tailored explicitly for graph data, ensuring that privacy is maintained without significantly compromising the accuracy of the learning algorithms [15].

Another emerging approach is the use of federated learning in graph analysis [78]. Federated learning allows multiple parties to train machine learning models collaboratively on decentralized data without sharing the raw data. In the context of graph learning, federated learning enables different organizations or entities to jointly analyze interconnected data while preserving the privacy of their individual datasets. This approach is precious in scenarios where data is distributed across different locations or owned by different stakeholders, such as in healthcare networks [6, 7, 124] or financial systems [18, 100].

DOI: 10.1201/9781003516613-9

9.2   Challenges in Scaling Privacy-Preserving Techniques

Scaling privacy-preserving techniques presents several significant challenges, particularly as data volumes and complexity continue to grow. One of the primary challenges is maintaining computational efficiency while implementing privacy measures. Privacy-preserving methods, such as differential privacy, homomorphic encryption, and secure multi-party computation, often introduce substantial computational overhead. As datasets expand and the complexity of tasks increases, the computational resources required to maintain privacy can become prohibitive, impacting the feasibility of these techniques for large-scale applications.

Another challenge is ensuring that privacy-preserving techniques can handle diverse and dynamic data environments. Real-world data is often heterogeneous and evolves over time, making it challenging to apply static privacy models effectively. Developing scalable solutions that can adapt to varying data types and structures while still ensuring robust privacy guarantees remains a complex problem. This challenge is exacerbated when dealing with real-time or streaming data, where privacy-preserving measures must be implemented swiftly and efficiently.

Furthermore, there are concerns related to the trade-offs between privacy and usability. As privacy-preserving techniques become more sophisticated, striking the right balance between privacy guarantees and the usability of the data or system becomes increasingly challenging. Ensuring that privacy measures do not overly degrade the quality or utility of the data is essential for maintaining the effectiveness of privacy-preserving solutions in practical scenarios.

9.3   Future Research Directions

Graph learning techniques for time-series data have been widely used in real applications, but there are still issues that need to be solved. Moreover, preserving the privacy of graph-structured data is an intricate and challenging undertaking. Therefore, future work can be focused on the following aspects:

- Dynamic or time-varying networks are of paramount importance in numerous real-world applications, and research endeavors to acquire valuable insights from such data form a crucial aspect of graph learning. In our future works, a primary focus will be devoted to the development of graph learning techniques that effectively capture and model the temporal dynamics inherent in graphs. Specifically, the endeavors can involve the design of dynamic graph networks capable of accurately representing and encapsulating the intricate changes in structure and evolving properties that characterize time-varying networks.

- Current work focused solely on graph learning from time series data, overlooking other important features that are relevant to graph clustering and community detection. Another focus of our future work is to expand the scope to explore graph learning techniques based on multi-feature data, such as the location and air quality in temperature data. Apart from inferring the graph topology, uncovering underlying patterns, similarities, and relationships among nodes based on their feature attributes can achieve graph clustering and community detection.

- DP provides a systematic and measurable framework for preserving privacy in data analysis. To ensure the privacy of latent graph structures and stimuli of graph-structured data, researchers are undertaking a thorough exploration and refinement of advanced methodologies building upon existing techniques. Upcoming endeavors involve conceptualizing and developing novel methods that integrate DP, reinforcing the privacy foundations of graph-structured data. In addition, there is a focus on the convergence of privacy and federated learning [183, 74]. In this area, innovative approaches could be explored to strike a delicate balance between safeguarding individual privacy within graph networks and leveraging the potential of federated learning for holistic insights.

10

Appendix

10.1 Appendix 1

The equivalence between the orthogonality requirement and the orthogonal projection is proved, as follows.

1) Proof of sufficiency condition: Suppose that $\mathbf{P}_{\mathbf{U}_{\mathcal{K}}} = \mathbf{U}_{\mathcal{K}}\mathbf{U}_{\mathcal{K}}^T$ is the orthogonal projection on the subspace spanned by the column-vectors of $\mathbf{U}_{\mathcal{K}}$. $\mathbf{P}_{\mathbf{U}_{\mathcal{K}}}\mathbf{U}_{\mathcal{K}} = \mathbf{U}_{\mathcal{K}}\mathbf{U}_{\mathcal{K}}^T\mathbf{U}_{\mathcal{K}} = \mathbf{U}_{\mathcal{K}}$. By multiplying $\mathbf{U}_{\mathcal{K}}^T$ to the left of both sides of $\mathbf{U}_{\mathcal{K}}\mathbf{U}_{\mathcal{K}}^T\mathbf{U}_{\mathcal{K}} = \mathbf{U}_{\mathcal{K}}$, it obtains

$$\mathbf{U}_{\mathcal{K}}^T\mathbf{U}_{\mathcal{K}}\mathbf{U}_{\mathcal{K}}^T\mathbf{U}_{\mathcal{K}} = \mathbf{U}_{\mathcal{K}}^T\mathbf{U}_{\mathcal{K}}. \tag{10.1}$$

Since $\mathbf{U}_{\mathcal{K}}$ is a full column rank matrix, $\mathbf{U}_{\mathcal{K}}^T\mathbf{U}_{\mathcal{K}}$ is a non-singular matrix. By multiplying $(\mathbf{U}_{\mathcal{K}}^T\mathbf{U}_{\mathcal{K}})^{-1}$ to the left of both sides of (10.1), it has

$$(\mathbf{U}_{\mathcal{K}}^T\mathbf{U}_{\mathcal{K}})^{-1}(\mathbf{U}_{\mathcal{K}}^T\mathbf{U}_{\mathcal{K}})(\mathbf{U}_{\mathcal{K}}^T\mathbf{U}_{\mathcal{K}}) = (\mathbf{U}_{\mathcal{K}}^T\mathbf{U}_{\mathcal{K}})^{-1}\mathbf{U}_{\mathcal{K}}^T\mathbf{U}_{\mathcal{K}}\mathbf{I}. \tag{10.2}$$

As a result, it concludes that $\mathbf{U}_{\mathcal{K}}^T\mathbf{U}_{\mathcal{K}} = \mathbf{I}$, i.e., $\mathbf{U}_{\mathcal{K}}$ is orthonormal.

2) Proof of necessary condition: Suppose that $\mathbf{U}_{\mathcal{K}}^T\mathbf{U}_{\mathcal{K}} = \mathbf{I}$. Based on the definition of orthogonal projection, it has $\mathbf{P}_{\mathbf{U}_{\mathcal{K}}} = \mathbf{U}_{\mathcal{K}}(\mathbf{U}_{\mathcal{K}}^T\mathbf{U}_{\mathcal{K}})^{-1}\mathbf{U}_{\mathcal{K}}^T = \mathbf{U}_{\mathcal{K}}\mathbf{U}_{\mathcal{K}}^T$. In other words, $P_{\mathbf{U}_{\mathcal{K}}}$ is the orthogonal projection on the subspace spanned by the column-vectors of $\mathbf{U}_{\mathcal{K}}$. In summary, the orthogonal requirement and the orthogonal projection are equivalent. The orthogonality requirement is preserved in (5.12) by using the orthogonal projection $\mathbf{P}_{\mathbf{U}_{\mathcal{K}}}$.

10.2 Appendix 2

Problem (5.15) is an unconstrained optimization problem that can be converted to a problem defined on a Grassmann manifold. The Grassmann manifold is a closed set, the maximum or minimum of a continuous function defined on the closed set, e.g., the optimal solution to (5.15), exists [120].

DOI: 10.1201/9781003516613-10

Suppose that the eigenvalues of $\left(\mathbf{I} - \mathbf{u}_1\mathbf{u}_1^T\right)\mathbf{Y}\mathbf{Y}^T\left(\mathbf{I} - \mathbf{u}_1\mathbf{u}_1^T\right)^T$ are $\sigma_1 \geq \sigma_2 \geq \cdots \geq \sigma_N$, corresponding to the eigenvectors $\{\mathbf{v}_1, \cdots, \mathbf{v}_N\}$. Let $S_1 =$ span $\{\mathbf{v}_1, \cdots, \mathbf{v}_K\}$, corresponding to the K largest eigenvalues $\{\sigma_1, \cdots, \sigma_K\}$. Let $S_2\,(S_2 \neq S_1)$ be any other K-dimensional subspace. Let $E_0 = S_1 \cap S_2$ and suppose that $S_1 = E_0 \oplus E_1$ and $S_2 = E_0 \oplus E_2$, where E_1 is the subset of S_1 and E_2 is the subset of $S_1^\perp$, i.e., $E_1 \subset$ span $\{\mathbf{v}_1, \cdots, \mathbf{v}_K\}$ and $E_2 \subset$ span $\{\mathbf{v}_{K+1}, \cdots, \mathbf{v}_N\}$; and $S_1^\perp$ is the orthogonal complement of S_1. Suppose that $\dim(E_1) = \dim(E_2) = t$. Based on the Minimax theorem,

$$\mathrm{tr}\left(\mathbf{P}_{E_1}\left(\mathbf{I} - \mathbf{u}_1\mathbf{u}_1^T\right)\mathbf{Y}\mathbf{Y}^T\left(\mathbf{I} - \mathbf{u}_1\mathbf{u}_1^T\right)^T\right) \geq t\sigma_K;$$
$$\mathrm{tr}\left(\mathbf{P}_{E_2}\left(\mathbf{I} - \mathbf{u}_1\mathbf{u}_1^T\right)\mathbf{Y}\mathbf{Y}^T\left(\mathbf{I} - \mathbf{u}_1\mathbf{u}_1^T\right)^T\right) \leq t\sigma_{K+1}. \tag{10.3}$$

As a result, it has

$$\mathrm{tr}\left(\mathbf{P}_{S_1}\left(\mathbf{I} - \mathbf{u}_1\mathbf{u}_1^T\right)\mathbf{Y}\mathbf{Y}^T\left(\mathbf{I} - \mathbf{u}_1\mathbf{u}_1^T\right)^T\right)$$
$$= \mathrm{tr}\left(\mathbf{P}_{E_0}\left(\mathbf{I} - \mathbf{u}_1\mathbf{u}_1^T\right)\mathbf{Y}\mathbf{Y}^T\left(\mathbf{I} - \mathbf{u}_1\mathbf{u}_1^T\right)^T\right) + \mathrm{tr}\left(\mathbf{P}_{E_1}\left(\mathbf{I} - \mathbf{u}_1\mathbf{u}_1^T\right)\mathbf{Y}\mathbf{Y}^T\left(\mathbf{I} - \mathbf{u}_1\mathbf{u}_1^T\right)^T\right)$$
$$\geq \mathrm{tr}\left(\mathbf{P}_{E_0}\left(\mathbf{I} - \mathbf{u}_1\mathbf{u}_1^T\right)\mathbf{Y}\mathbf{Y}^T\left(\mathbf{I} - \mathbf{u}_1\mathbf{u}_1^T\right)^T\right) + \mathrm{tr}\left(\mathbf{P}_{E_2}\left(\mathbf{I} - \mathbf{u}_1\mathbf{u}_1^T\right)\mathbf{Y}\mathbf{Y}^T\left(\mathbf{I} - \mathbf{u}_1\mathbf{u}_1^T\right)^T\right)$$
$$= \mathrm{tr}\left(\mathbf{P}_{S_2}\left(\mathbf{I} - \mathbf{u}_1\mathbf{u}_1^T\right)\mathbf{Y}\mathbf{Y}^T\left(\mathbf{I} - \mathbf{u}_1\mathbf{u}_1^T\right)^T\right), \tag{10.4a}$$

where the two equations are based on $S_1 = E_0 \oplus E_1$ and $S_2 = E_0 \oplus E_2$. Hence, the projection of $\left(\mathbf{I} - \mathbf{u}_1\mathbf{u}_1^T\right)\mathbf{Y}\mathbf{Y}^T\left(\mathbf{I} - \mathbf{u}_1\mathbf{u}_1^T\right)$ is the largest on the span associated with the K largest eigenvalues. The solution to (5.15) is $\mathbf{U}_K^* =$ span $\{\mathbf{v}_1, \cdots, \mathbf{v}_K\}$, i.e., the K largest eigenvalues of $\left(\mathbf{I} - \mathbf{u}_1\mathbf{u}_1^T\right)\mathbf{Y}\mathbf{Y}^T\left(\mathbf{I} - \mathbf{u}_1\mathbf{u}_1^T\right)^T$.

10.3   Appendix 3

ADMM is an augmented Lagrangian method. With the step size ρ (e.g., $\rho = 1$ by default in Section 5.6), the augmented Lagrangian of (5.26) is given by

$$\mathcal{L}\left(\mathbf{\Lambda}_1, \mathbf{\Lambda}_2, \mathbf{C}, \mathbf{Z}\right) = \mathrm{tr}\left(\mathbf{\Lambda}_1\widetilde{\mathbf{T}}_1\right) + \mathrm{tr}\left(\mathbf{\Lambda}_2\widetilde{\mathbf{T}}_2\right) - M\log\det\left(\mathbf{\Lambda}_1\right) \tag{10.5a}$$
$$- M\log\det\left(\mathbf{\Lambda}_2\right) + \mathrm{tr}\left(\mathbf{Z}^T\left(\mathbf{V}_1\mathbf{\Lambda}_1\mathbf{V}_1^T + \mathbf{V}_2\mathbf{\Lambda}_2\mathbf{V}_2^T - \mathbf{C}\right)\right) \tag{10.5b}$$
$$+ \frac{\rho}{2}\left\|\mathbf{V}_1\mathbf{\Lambda}_1\mathbf{V}_1^T + \mathbf{V}_2\mathbf{\Lambda}_2\mathbf{V}_2^T - \mathbf{C}\right\|_F^2, \tag{10.5c}$$

where $\mathbf{Z}$ is the dual variable associated with (5.26c).

Following the standard ADMM framework, the primal variables, $\mathbf{\Lambda}_1$, $\mathbf{\Lambda}_2$, and $\mathbf{C}$, and the dual variable $\mathbf{Z}$ are updated in an alternating manner:

Update $\mathbf{\Lambda}_1$ *and* $\mathbf{\Lambda}_2$: Based on $\mathbf{\Lambda}_2^l$, $\mathbf{C}^l$, and $\mathbf{Z}^l$ obtained in the l-th iteration, $\mathbf{\Lambda}_1^{l+1}$ in the $(l+1)$-th iteration can be obtained by

$$\mathbf{\Lambda}_1^{l+1} = \arg\min_{\mathbf{\Lambda}_1 \succeq 0} \mathcal{L}\left(\mathbf{\Lambda}_1, \mathbf{\Lambda}_2^l, \mathbf{C}^l, \mathbf{Z}^l\right) \tag{10.6}$$

$$= \arg\min_{\mathbf{\Lambda}_1 \succeq 0}\left\{\frac{\rho}{2}\left\|\mathbf{\Lambda}_1 + \frac{1}{\rho}(\widetilde{\mathbf{T}}_1 + \widetilde{\mathbf{Z}}_1 - \rho\widetilde{\mathbf{X}}_1)\right\|_F^2 - M\log\det\left(\mathbf{\Lambda}_1^l\right)\right\},$$

where $\widetilde{\mathbf{Z}}_1 = \mathbf{V}_1^T(\mathbf{Z}^l)^T\mathbf{V}_1$ and $\widetilde{\mathbf{X}}_1 = \mathbf{V}_1^T(\mathbf{C}^l - \mathbf{V}_2\mathbf{\Lambda}_2^l\mathbf{V}_2^T)\mathbf{V}_1$. By setting the first-order derivative of $\mathcal{L}\left(\mathbf{\Lambda}_1, \mathbf{\Lambda}_2^l, \mathbf{C}^l, \mathbf{Z}^l\right)$ with respect to $\mathbf{\Lambda}_1$ to $\mathbf{0}$, it has

$$\rho\mathbf{\Lambda}_1 - M(\mathbf{\Lambda}_1)^\dagger = -(\widetilde{\mathbf{T}}_1 + \widetilde{\mathbf{Z}}_1 - \rho\widetilde{\mathbf{X}}_1). \tag{10.7}$$

Taking the orthogonal eigenvalue decomposition of the right-hand side of (10.7) yields $\rho\mathbf{\Lambda}_1 - M(\mathbf{\Lambda}_1)^\dagger = \mathbf{Q}_1\mathbf{\Xi}_1\mathbf{Q}_1^T$. Then, by left multiplying $\mathbf{Q}_1^T$ and right multiplying $\mathbf{Q}_1$ on both sides, it yields $\rho\widetilde{\mathbf{\Lambda}}_1 - M(\widetilde{\mathbf{\Lambda}}_1)^\dagger = \mathbf{\Xi}_1$. $\widetilde{\mathbf{\Lambda}}_1 = \mathbf{Q}_1^T\mathbf{\Lambda}_1\mathbf{Q}_1$. Here, $\mathbf{Q}_1$ is the unitary matrix with $\mathbf{Q}_1\mathbf{Q}_1^T = \mathbf{I}$. The diagonal matrix $\mathbf{\Xi}_1 = \text{diag}(\xi_{11}, \cdots, \xi_{1K})$ collects the eigenvalues of $(\widetilde{\mathbf{T}}_1 + \widetilde{\mathbf{Z}}_1 - \rho\widetilde{\mathbf{X}}_1)$. Using the quadratic formula, the non-negative solution to $\rho\widetilde{\mathbf{\Lambda}}_1 - M(\widetilde{\mathbf{\Lambda}}_1)^\dagger = \mathbf{\Xi}_1$ is $\widetilde{\Lambda}_{1i} = \frac{-\rho\xi_{1i} + \sqrt{\rho^2\xi_{1i}^2 + 4M\rho}}{2\rho}$. With $\widetilde{\mathbf{\Lambda}}_1 = \text{diag}(\widetilde{\Lambda}_{11}, \cdots, \widetilde{\Lambda}_{1K})$, the solution to (10.6) is

$$\mathbf{\Lambda}_1^{l+1} = \mathbf{Q}_1\widetilde{\mathbf{\Lambda}}_1\mathbf{Q}_1^T, \tag{10.8}$$

Given $\mathbf{\Lambda}_1^{l+1}$, the primal variable $\mathbf{\Lambda}_2^{l+1}$ can be obtained in the same way.

Update $\mathbf{C}$: Given $\mathbf{\Lambda}_1^{l+1}$ and $\mathbf{\Lambda}_2^{l+1}$ in the $(l+1)$-th iteration, it obtains

$$\mathbf{C}^{l+1} = \arg\min_{\mathbf{C}} \mathcal{L}\left(\mathbf{\Lambda}_1^{l+1}, \mathbf{\Lambda}_2^{l+1}, \mathbf{C}^l, \mathbf{Z}^l\right) =$$

$$\arg\min_{\mathbf{C}}\left\{\frac{\rho}{2}\left\|\mathbf{C}^l - \left(\frac{1}{\rho}(\mathbf{Z}^l)^T + \mathbf{V}_1\mathbf{\Lambda}_1^{l+1}\mathbf{V}_1^T + \mathbf{V}_2\mathbf{\Lambda}_2^{l+1}\mathbf{V}_2^T\right)\right\|_F^2\right\}. \tag{10.9}$$

Consider the constraints $\mathbf{I} \odot \mathbf{C} \geq 0$ and $\mathbf{A} \odot \mathbf{C} \leq 0$ in (5.26); the diagonal elements are non-negative and the off-diagonal elements are non-positive. The solution to (10.9) is obtained:

$$\mathbf{C}^{l+1} = \mathbf{I} \odot \left[\frac{1}{\rho}(\mathbf{Z}^l)^T + \mathbf{V}_1\mathbf{\Lambda}_1^{l+1}\mathbf{V}_1^T + \mathbf{V}_2\mathbf{\Lambda}_2^{l+1}\mathbf{V}_2^T\right]_+$$

$$+ \mathbf{A} \odot \left[\frac{1}{\rho}(\mathbf{Z}^l)^T + \mathbf{V}_1\mathbf{\Lambda}_1^{l+1}\mathbf{V}_1^T + \mathbf{V}_2\mathbf{\Lambda}_2^{l+1}\mathbf{V}_2^T\right]_-. \tag{10.10}$$

Update $\mathbf{Z}$: Given $\mathbf{\Lambda}_1^{l+1}$, $\mathbf{\Lambda}_2^{l+1}$ and $\mathbf{C}^{l+1}$, then $\mathbf{Z}^{l+1}$ can be obtained by

$$\mathbf{Z}^{l+1} = \mathbf{Z}^l + \rho\left(\mathbf{V}_1\mathbf{\Lambda}_1^{l+1}\mathbf{V}_1^T + \mathbf{V}_2\mathbf{\Lambda}_2^{l+1}\mathbf{V}_2^T - \mathbf{C}^{l+1}\right). \tag{10.11}$$

The convergence criterion of the ADMM is $\|\mathbf{C}^{l+1} - \mathbf{C}^l\|/\|\mathbf{C}^l\| < \xi$ and $\|\mathbf{Z}^{l+1} - \mathbf{Z}^l\|/\|\mathbf{Z}^l\| < \xi$, where ξ is a predefined threshold.

Bibliography

[1] Mehran Abolhasan, Justin Lipman, Negin Shariati, Wei Ni, Abbas Jamalipour, et al. Joint mobile node participation and multihop routing for emerging open radio-based intelligent transportation system. *IEEE Access*, 10:85228–85242, 2022.

[2] Nurul Absar, Baitul Mamur, Abir Mahmud, Talha Bin Emran, Mayeen Uddin Khandaker, MRI Faruque, Hamid Osman, Amin Elzaki, and Bahaaedin A Elkhader. Development of a computer-aided tool for detection of Covid-19 pneumonia from CXR images using machine learning algorithm. *JRRAS*, 15(1):32–43, 2022.

[3] Priya Aggarwal and Anubha Gupta. Low rank and sparsity constrained method for identifying overlapping functional brain networks. *Plos one*, 13(11):e0208068, 2018.

[4] Kimberly M Albert, Guy G Potter, Brian D Boyd, Hakmook Kang, and Warren D Taylor. Brain network functional connectivity and cognitive performance in major depressive disorder. *J. Psychiatr. Res.*, 110:51–56, 2019.

[5] Rasim Alguliyev, Ramiz Aliguliyev, and Farhad Yusifov. Graph modelling for tracking the Covid-19 pandemic spread. *Infect. Dis. Model.*, 6:112–122, 2021.

[6] Mansoor Ali, Faisal Naeem, Muhammad Tariq, and Georges Kaddoum. Federated learning for privacy preservation in smart healthcare systems: A comprehensive survey. *IEEE journal of biomedical and health informatics*, 27(2):778–789, 2022.

[7] Rodolfo Stoffel Antunes, Cristiano André da Costa, Arne Küderle, Imrana Abdullahi Yari, and Björn Eskofier. Federated learning for healthcare: Systematic review and architecture proposal. *ACM Transactions on Intelligent Systems and Technology (TIST)*, 13(4):1–23, 2022.

[8] Giuseppe Arbia, Riccardo Bramante, Silvia Facchinetti, and Diego Zappa. Modeling inter-country spatial financial interactions with graphical lasso: An application to sovereign co-risk evaluation. *Regional Science and Urban Economics*, 70:72–79, 2018.

[9] Amir R Asadi, Emmanuel Abbe, and Sergio Verdú. Compressing data on graphs with clusters. In *Proc. ISIT*, pages 1583–1587. IEEE, 2017.

[10] Sarita Azad and Sushma Devi. Tracking the spread of Covid-19 in India via social networks in the early phase of the pandemic. *J. Travel Med.*, 27(8):taaa130, 2020.

[11] Albert-László Barabási and Réka Albert. Emergence of scaling in random networks. *Science*, 286(5439):509–512, 1999.

[12] Richard G Baraniuk, Volkan Cevher, Marco F Duarte, and Chinmay Hegde. Model-based compressive sensing. *IEEE Trans. Inf. Theory*, 56(4):1982–2001, 2010.

[13] David J Bartholomew, Martin Knott, and Irini Moustaki. *Latent variable models and factor analysis: A unified approach*. John Wiley & Sons, 2011.

[14] Kamal Berahmand, Elahe Nasiri, Mehrdad Rostami, and Saman Forouzandeh. A modified deepwalk method for link prediction in attributed social network. *Computing*, 103:2227–2249, 2021.

[15] Karuna Bhaila, Wen Huang, Yongkai Wu, and Xintao Wu. Local differential privacy in graph neural networks: a reconstruction approach. In *Proceedings of the 2024 SIAM International Conference on Data Mining (SDM)*, pages 1–9. SIAM, 2024.

[16] Ulrik Brandes. A faster algorithm for betweenness centrality. *J. Math. Sociol.*, 25(2):163–177, 2001.

[17] Michael M Bronstein, Joan Bruna, Yann LeCun, Arthur Szlam, and Pierre Vandergheynst. Geometric deep learning: Going beyond Euclidean data. *IEEE Signal Process. Mag.*, 34(4):18–42, 2017.

[18] David Byrd and Antigoni Polychroniadou. Differentially private secure multi-party computation for federated learning in financial applications. In *Proceedings of the First ACM International Conference on AI in Finance*, pages 1–9, 2020.

[19] Qingjiu Cao, Ni Shu, Li An, Peng Wang, Li Sun, Ming-Rui Xia, et al. Probabilistic diffusion tractography and graph theory analysis reveal abnormal white matter structural connectivity networks in drug-naive boys with attention deficit/hyperactivity disorder. *Journal of Neuroscience*, 33(26):10676–10687, 2013.

[20] Wei Koong Chai. Modelling spreading process induced by agent mobility in complex networks. *IEEE Trans. Netw. Sci. Eng.*, 5(4):336–349, 2017.

[21] Siheng Chen, Aliaksei Sandryhaila, José MF Moura, and Jelena Kovačević. Signal recovery on graphs: Variation minimization. *IEEE Trans. Signal Process.*, 63(17):4609–4624, 2015.

[22] Siheng Chen, Aliaksei Sandryhaila, José MF Moura, and Jelena Kovačević. Signal recovery on graphs: Variation minimization. *IEEE Trans. Signal Process.*, 63(17):4609–4624, 2015.

[23] Siheng Chen, Rohan Varma, Aliaksei Sandryhaila, and Jelena Kovačević. Discrete signal processing on graphs: Sampling theory. *IEEE Trans. Signal Process.*, 63(24):6510–6523, 2015.

[24] Xiaojing Chen, Hanfei Wen, Wei Ni, Shunqing Zhang, Xin Wang, Shugong Xu, and Qingqi Pei. Distributed online optimization of edge computing with mixed power supply of renewable energy and smart grid. *IEEE Trans. Commun.*, 70(1):389–403, 2021.

[25] Xu Chen, Zhiyong Feng, J Andrew Zhang, Feifei Gao, Xin Yuan, Zhaohui Yang, and Ping Zhang. Complex CNN CSI enhancer for integrated sensing and communications. *IEEE Journal of Selected Topics in Signal Processing*, 2024.

[26] Xu Chen, Zhiyong Feng, J Andrew Zhang, Xin Yuan, and Ping Zhang. Integrated sensing and communication complex CNN CSI enhancer for 6g networks. *arXiv preprint arXiv:2305.17938*, 2023.

[27] Sundeep Prabhakar Chepuri and Geert Leus. Graph sampling for covariance estimation. *IEEE Trans. Signal Inf. Process. Netw.*, 3(3):451–466, 2017.

[28] Sundeep Prabhakar Chepuri, Sijia Liu, Geert Leus, and Alfred O Hero. Learning sparse graphs under smoothness prior. In *Proc. ICASSP*, pages 6508–6512. IEEE, 2017.

[29] Amanda MY Chu, Jacky NL Chan, Jenny TY Tsang, Agnes Tiwari, and Mike KP So. Analyzing cross-country pandemic connectedness during Covid-19 using a spatial-temporal database: Network analysis. *JMIR Public Health Surveill.*, 7(3):e27317, 2021.

[30] Amanda MY Chu, Thomas WC Chan, Mike KP So, and Wing-Keung Wong. Dynamic network analysis of Covid-19 with a latent pandemic space model. *Int. J. Environ. Res. Public Health*, 18(6):3195, 2021.

[31] Mario Coutino, Elvin Isufi, and Geert Leus. Advances in distributed graph filtering. *IEEE Trans. Signal Process.*, 67(9):2320–2333, 2019.

[32] Pierluigi Crescenzi, Gianlorenzo D'angelo, Lorenzo Severini, and Yllka Velaj. Greedily improving our own closeness centrality in a network. *ACM Trans. Knowl. Discov. Data*, 11(1):1–32, 2016.

[33] Qimei Cui, Xingxing Hu, Wei Ni, Xiaofeng Tao, Ping Zhang, Tao Chen, Kwang-Cheng Chen, and Martin Haenggi. Vehicular mobility patterns and their applications to internet-of-vehicles: A comprehensive survey. *Sci. China Inf. Sci.*, 65(11):1–42, 2022.

[34] Qimei Cui, Wei Ni, Shenghong Li, Borui Zhao, Ren Ping Liu, and Ping Zhang. Learning-assisted clustered access of 5g/b5g networks to unlicensed spectrum. *IEEE Wirel. Commun.*, 27(1):31–37, 2020.

[35] Qimei Cui, Yingze Wang, Kwang-Cheng Chen, Wei Ni, I-Cheng Lin, Xiaofeng Tao, and Ping Zhang. Big data analytics and network calculus enabling intelligent management of autonomous vehicles in a smart city. *IEEE Internet Things J.*, 6(2):2021–2034, 2018.

[36] Jingchao Dai, Keke Huang, Yishun Liu, Chunhua Yang, and Zhen Wang. Global reconstruction of complex network topology via structured compressive sensing. *IEEE Syst. J.*, 15(2):1959–1969, 2020.

[37] Jesper Dall and Michael Christensen. Random geometric graphs. *Physical review E*, 66(1):016121, 2002.

[38] Wei-Yen Day, Ninghui Li, and Min Lyu. Publishing graph degree distribution with node differential privacy. In *Proceedings of the 2016 International Conference on Management of Data*, pages 123–138, 2016.

[39] Konstantinos Demertzis, Dimitrios Tsiotas, and Lykourgos Magafas. Modeling and forecasting the Covid-19 temporal spread in Greece: An exploratory approach based on complex network defined splines. *Int. J. Environ. Res. Public Health*, 17(13):4693, 2020.

[40] Arthur P Dempster. Covariance selection. *Biometrics*, pages 157–175, 1972.

[41] Yu Deng, Yahua Zhang, and Kun Wang. An analysis of the Chinese scheduled freighter network during the first year of the Covid-19 pandemic. *J. Transp. Geogr.*, 99:103298, 2022.

[42] Paolo Di Lorenzo et al. Adaptive graph signal processing: Algorithms and optimal sampling strategies. *IEEE Trans. Signal Process.*, 66(13):3584–3598, 2018.

[43] Xiaofeng Ding, Cui Wang, Kim-Kwang Raymond Choo, and Hai Jin. A novel privacy preserving framework for large scale graph data publishing. *IEEE Trans. Knowl. Data Eng.*, 33(2):331–343, 2019.

[44] Xiaowen Dong, Dorina Thanou, Pascal Frossard, and Pierre Vandergheynst. Learning Laplacian matrix in smooth graph signal representations. *IEEE Trans. Signal Process.*, 64(23):6160–6173, 2016.

[45] Xiaowen Dong, Dorina Thanou, Michael Rabbat, and Pascal Frossard. Learning graphs from data: A signal representation perspective. *IEEE Signal Process. Mag.*, 36(3):44–63, 2019.

[46] Daiana Caroline dos Santos Gomes and Ginalber Luiz de Oliveira Serra. Machine learning model for computational tracking and forecasting the

Covid-19 dynamic propagation. *IEEE J. Biomed. Health informat.*, 25(3):615–622, 2021.

[47] Kuaikuai Duan, Wenhao Jiang, Kelly Rootes-Murdy, Gido H Schoenmacker, Alejandro Arias-Vasquez, Jan K Buitelaar, Martine Hoogman, Jaap Oosterlaan, Pieter J Hoekstra, Dirk J Heslenfeld, et al. Gray matter networks associated with attention and working memory deficit in ADHD across adolescence and adulthood. *Translational psychiatry*, 11(1):1–12, 2021.

[48] Cynthia Dwork, Aaron Roth, et al. The algorithmic foundations of differential privacy. *Foundations and Trends® in Theoretical Computer Science*, 9(3–4):211–407, 2014.

[49] Hana-May Eadeh, Kristian E Markon, Joel T Nigg, and Molly A Nikolas. Evaluating the viability of neurocognition as a transdiagnostic construct using both latent variable models and network analysis. *Research on Child and Adolescent Psychopathology*, 49(6):697–710, 2021.

[50] Richard eare, Chris Adamson, Mark A Bellgrove, Veronika Vilgis, Alasdair Vance, et al. Altered structural connectivity in ADHD: A network based analysis. *Brain Imag Behav.*, 11(3):846–858, 2017.

[51] Hilmi E Egilmez, Yung-Hsuan Chao, and Antonio Ortega. Graph-based transforms for video coding. *IEEE Trans. Image Process.*, 29:9330–9344, 2020.

[52] Hilmi E Egilmez, Eduardo Pavez, and Antonio Ortega. Graph learning from data under Laplacian and structural constraints. *IEEE J. Sel. Topics Signal Process.*, 11(6):825–841, 2017.

[53] Yonina C Eldar and Moshe Mishali. Robust recovery of signals from a structured union of subspaces. *IEEE Trans. Inf. Theory*, 55(11):5302–5316, 2009.

[54] Yousef Emami, Bo Wei, Kai Li, Wei Ni, and Eduardo Tovar. Joint communication scheduling and velocity control in multi-UAV-assisted sensor networks: A deep reinforcement learning approach. *IEEE Trans. Veh. Technol.*, 70(10):10986–10998, 2021.

[55] Paul Erdős and Alfréd Rényi. On the evolution of random graphs. *Publ. Math. Inst. Hung. Acad. Sci*, 5(1):17–60, 1960.

[56] Stephen V Faraone, Tobias Banaschewski, David Coghill, Yi Zheng, Joseph Biederman, Mark A Bellgrove, Jeffrey H Newcorn, Martin Gignac, Nouf M Al Saud, Iris Manor, et al. The world federation of ADHD international consensus statement: 208 evidence-based conclusions about the disorder. *Neurosci. Biobehav. Rev.*, 128:789–818, 2021.

[57] Zhiming Feng, Chiwei Xiao, Peng Li, Zhen You, Xu Yin, and Fangyu Zheng. Comparison of spatio-temporal transmission characteristics of Covid-19 and its mitigation strategies in China and the US. *J. Geogr. Sci.*, 30(12):1963–1984, 2020.

[58] Pau Ferrer-Cid et al. Graph learning techniques using structured data for IoT air pollution monitoring platforms. *IEEE Internet Things J.*, 8(17):13652–13663, 2021.

[59] Massimo Filippi, Silvia Basaia, Elisa Canu, Francesca Imperiale, Giuseppe Magnani, Monica Falautano, Giancarlo Comi, Andrea Falini, and Federica Agosta. Changes in functional and structural brain connectome along the alzheimer's disease continuum. *Molecular psychiatry*, 25(1):230–239, 2020.

[60] Jerome Friedman, Trevor Hastie, and Robert Tibshirani. Sparse inverse covariance estimation with the graphical lasso. *Biostatistics*, 9(3):432–441, 2008.

[61] Karl J Friston, Andrew P Holmes, JB Poline, PJ Grasby, SCR Williams, Richard SJ Frackowiak, and Robert Turner. Analysis of fMRI time-series revisited. *Neuroimage*, 2(1):45–53, 1995.

[62] Siyuan Gao, Xinyue Xia, Dustin Scheinost, and Gal Mishne. Smooth graph learning for functional connectivity estimation. *NeuroImage*, 239:118289, 2021.

[63] Georgios B Giannakis, Yanning Shen, and Georgios Vasileios Karaniko-las. Topology identification and learning over graphs: Accounting for nonlinearities and dynamics. *Proc. IEEE*, 106(5):787–807, 2018.

[64] Yudong Gong, Sanyang Liu, and Yiguang Bai. Efficient parallel computing on the game theory-aware robust influence maximization problem. *Knowl. Based Syst.*, 220:106942, 2021.

[65] Jun-ya Gotoh, Akiko Takeda, and Katsuya Tono. DC formulations and algorithms for sparse optimization problems. *Mathematical Programming*, 169(1):141–176, 2018.

[66] Guangyu Guo, Zhuoyan Liu, Shijie Zhao, Lei Guo, and Tianming Liu. Eliminating indefiniteness of clinical spectrum for better screening Covid-19. *IEEE J. Biomed. Health informat.*, 25(5):1347–1357, 2021.

[67] David K Hammond, Pierre Vandergheynst, and Rémi Gribonval. Wavelets on graphs via spectral graph theory. *Applied and Computational Harmonic Analysis*, 30(2):129–150, 2011.

[68] Yi Han, Lan Yang, Kun Jia, Jie Li, Siyuan Feng, Wei Chen, Wenwu Zhao, and Paulo Pereira. Spatial distribution characteristics of the

Covid-19 pandemic in Beijing and its relationship with environmental factors. *Sci. Total Environ.*, 761:144257, 2021.

[69] Michael Hay, Gerome Miklau, David Jensen, Philipp Weis, and Siddharth Srivastava. Anonymizing social networks. *Computer science department faculty publication series*, page 180, 2007.

[70] Jianping He, Lin Cai, Chengcheng Zhao, Peng Cheng, and Xinping Guan. Privacy-preserving average consensus: privacy analysis and algorithm design. *IEEE Trans. Signal Inform. Process. Netw.*, 5(1):127–138, 2018.

[71] Soon-Beom Hong, Andrew Zalesky, Alex Fornito, Subin Park, Young-Hui Yang, Min-Hyeon Park, et al. Connectomic disturbances in attention-deficit/hyperactivity disorder: A whole-brain tractography analysis. *Biological psychiatry*, 76(8):656–663, 2014.

[72] Chenhui Hu, Sepulcre Jorge, A Keith, Johnson, E Georges, Fakhri, M Yue, and Lu. Matched signal detection on graphs: Theory and application to brain imaging data classification. *NeuroImage*, 125:587–600, 2016.

[73] Shuyan Hu, Xiaojing Chen, Wei Ni, Ekram Hossain, and Xin Wang. Distributed machine learning for wireless communication networks: Techniques, architectures, and applications. *IEEE Commun. Surv. Tutor.*, 23(3):1458–1493, 2021.

[74] Shuyan Hu, Xin Yuan, Wei Ni, Xin Wang, Ekram Hossain, and H Vincent Poor. Ofdma-f 2 1: Federated learning with flexible aggregation over an OFDMA air interface. *IEEE Trans. Wirel. Commun.*, 2024.

[75] Haiping Huang, Dongjun Zhang, Fu Xiao, Kai Wang, Jiateng Gu, and Ruchuan Wang. Privacy-preserving approach PBCN in social network with differential privacy. *IEEE Trans. Netw. Service Manag.*, 17(2):931–945, 2020.

[76] Huaxi Huang, Junjie Zhang, Litao Yu, Jian Zhang, Qiang Wu, and Chang Xu. Toan: Target-oriented alignment network for fine-grained image categorization with few labeled samples. *IEEE Trans. Circuits Syst. Video Technol.*, 32(2):853–866, 2021.

[77] Shuai Huang, Jing Li, Liang Sun, Jun Liu, Teresa Wu, Kewei Chen, Adam Fleisher, Eric Reiman, and Jieping Ye. Learning brain connectivity of alzheimer's disease from neuroimaging data. *Advances in Neural Information Processing Systems*, 22, 2009.

[78] Wenke Huang, Guancheng Wan, Mang Ye, and Bo Du. Federated graph semantic and structural learning. *arXiv preprint arXiv:2406.18937*, 2024.

[79] Pierre Humbert, Batiste Le Bars, Laurent Oudre, Argyris Kalogeratos, and Nicolas Vayatis. Learning Laplacian matrix from graph signals with sparse spectral representation. Technical report, 2019.

[80] Pierre Humbert, Batiste Le Bars, Laurent Oudre, Argyris Kalogeratos, and Nicolas Vayatis. Learning Laplacian matrix from graph signals with sparse spectral representation. *J. Mach. Learn. Res.*, 22(195):1–47, 2021.

[81] Cosimo Ieracitano, Nadia Mammone, Mario Versaci, Giuseppe Varone, Abder-Rahman Ali, Antonio Armentano, Grazia Calabrese, Anna Ferrarelli, Lorena Turano, Carmela Tebala, et al. A fuzzy-enhanced deep learning approach for early detection of Covid-19 pneumonia from portable chest x-ray images. *Neurocomputing*, 481:202–215, 2022.

[82] Ireneusz Jabłoński. Graph signal processing in applications to sensor networks, smart grids, and smart cities. *IEEE Sens. J.*, 17(23):7659–7666, 2017.

[83] Ankit Kumar Jain, Somya Ranjan Sahoo, and Jyoti Kaubiyal. Online social networks security and privacy: Comprehensive review and analysis. *Complex & Intelligent Systems*, 7(5):2157–2177, 2021.

[84] Di Jin, Xiaobao Wang, Mengquan Liu, Jianguo Wei, Wenhuan Lu, et al. Identification of generalized semantic communities in large social networks. *IEEE Trans. Netw. Sci. Eng.*, 7(4):2966–2979, 2020.

[85] Wonkwang Jo, Dukjin Chang, Myoungsoon You, and Ghi-Hoon Ghim. A social network analysis of the spread of Covid-19 in South Korea and policy implications. *Sci. Rep.*, 11(1):1–10, 2021.

[86] Vassilis Kalofolias. How to learn a graph from smooth signals. In *Proc. 19th Int. Conf. Artif. Intell. Statist., Cadiz, Spain,*, pages 920–929, 2016.

[87] Zhao Kang, Zhiping Lin, Xiaofeng Zhu, and Wenbo Xu. Structured graph learning for scalable subspace clustering: From single view to multiview. *IEEE Trans. Cybern.*, 2021.

[88] Nima Kianfar, Mohammad Saadi Mesgari, Abolfazl Mollalo, and Mehrdad Kaveh. Spatio-temporal modeling of Covid-19 prevalence and mortality using artificial neural network algorithms. *Spatial and Spatio-temporal Epidemiology*, 40:100471, 2022.

[89] Aaron Kucyi, Arielle Tambini, Sepideh Sadaghiani, Shella Keilholz, and Jessica R Cohen. Spontaneous cognitive processes and the behavioral validation of time-varying brain connectivity. *Netw. Neurosci.*, 2(4):397–417, 2018.

[90] Askat Kuzdeuov, Daulet Baimukashev, Aknur Karabay, Bauyrzhan Ibragimov, Almas Mirzakhmetov, Mukhamet Nurpeiissov, Michael Lewis, and Huseyin Atakan Varol. A network-based stochastic epidemic

simulator: Controlling Covid-19 with region-specific policies. *IEEE J. Biomed. Health informat.*, 24(10):2743–2754, 2020.

[91] Batiste Le Bars, Pierre Humbert, Laurent Oudre, and Argyris Kalogeratos. Learning Laplacian matrix from bandlimited graph signals. In *Proc. Int. Conf. Acoust., Speech, Signal Process.*, pages 2937–2941. IEEE, 2019.

[92] Yann LeCun, Yoshua Bengio, and Geoffrey Hinton. Deep learning. *nature*, 521(7553):436–444, 2015.

[93] Dandan Li, Ting Li, Yan Niu, Jie Xiang, Rui Cao, BO Liu, Hui Zhang, and Bin Wang. Reduced hemispheric asymmetry of brain anatomical networks in attention deficit hyperactivity disorder. *Brain imaging and behavior*, 13(3):669–684, 2019.

[94] Kai Li, Qimei Cui, Zengbao Zhu, Wei Ni, and Xiaofeng Tao. Lightweight, privacy-preserving handover authentication for integrated terrestrial-satellite networks. In *ICC 2022-IEEE International Conference on Communications*, pages 25–31. IEEE, 2022.

[95] Kai Li, Wei Ni, Harrison Kurunathan, and Falko Dressler. Data-driven deep reinforcement learning for online flight resource allocation in uav-aided wireless powered sensor networks. In *ICC 2022-IEEE International Conference on Communications*, pages 1–6. IEEE, 2022.

[96] Kai Li, Razvan Christian Voicu, Salil S Kanhere, Wei Ni, and Eduardo Tovar. Energy efficient legitimate wireless surveillance of UAV communications. *IEEE Trans. Veh. Technol.*, 68(3):2283–2293, 2019.

[97] Kai Li, Xin Yuan, Jingjing Zheng, Wei Ni, Falko Dressler, and Abbas Jamalipour. Leverage variational graph representation for model poisoning on federated learning. *IEEE Trans. Neural Netwo. Learn. Syst.*, pages 1–13, 2024.

[98] Kai Li, Xin Yuan, Jingjing Zheng, Wei Ni, and Mohsen Guizani. Exploring adversarial graph autoencoders to manipulate federated learning in the internet of things. In *2023 International Wireless Communications and Mobile Computing (IWCMC)*, pages 898–903, 2023.

[99] Kai Li, Jingjing Zheng, Xin Yuan, Wei Ni, Ozgur B. Akan, and H. Vincent Poor. Data-agnostic model poisoning against federated learning: A graph autoencoder approach. *IEEE Trans. Inf. Forensics Secur.*, 19:3465–3480, 2024.

[100] Qinbin Li, Zeyi Wen, Zhaomin Wu, Sixu Hu, Naibo Wang, Yuan Li, Xu Liu, and Bingsheng He. A survey on federated learning systems: Vision, hype and reality for data privacy and protection. *IEEE Transactions on Knowledge and Data Engineering*, 35(4):3347–3366, 2021.

[101] Qiongxiu Li, Mario Coutino, Geert Leus, and Mads Græsbøll Christensen. Privacy-preserving distributed graph filtering. In *2020 28th European Signal Processing Conference (EUSIPCO)*, pages 2155–2159. IEEE, 2021.

[102] Qiongxiu Li, Richard Heusdens, and Mads Græsbøll Christensen. Privacy-preserving distributed optimization via subspace perturbation: A general framework. *IEEE Trans. Signal Process.*, 68:5983–5996, 2020.

[103] Ren-Cang Li. Matrix perturbation theory. In *Handbook of linear algebra*, pages 15–1. Chapman and Hall/CRC, 2006.

[104] Rui Li, Xin Yuan, Mohsen Radfar, Peter Marendy, Wei Ni, Terence J O'Brien, and Pablo M Casillas-Espinosa. Graph signal processing, graph neural network and graph learning on biological data: A systematic review. *IEEE Rev. Biomed. Eng.*, 2021.

[105] Xiang Li, Lingyun Lu, Wei Ni, Abbas Jamalipour, Dalin Zhang, and Haifeng Du. Federated multi-agent deep reinforcement learning for resource allocation of vehicle-to-vehicle communications. *IEEE Trans. Veh. Technol.*, 71(8):8810–8824, 2022.

[106] Xiangjun Li, Qimei Cui, Qiulin Xue, Wei Ni, Jing Guo, and Xiaofeng Tao. A new batch access scheme with global qos optimization for satellite-terrestrial networks. In *GLOBECOM 2022-2022 IEEE Global Communications Conference*, pages 3929–3934. IEEE, 2022.

[107] Xiaoxiao Li, Yuan Zhou, Nicha Dvornek, Muhan Zhang, Siyuan Gao, Juntang Zhuang, Dustin Scheinost, Lawrence H Staib, Pamela Ventola, and James S Duncan. Braingnn: Interpretable brain graph neural network for fMRI analysis. *Medical Image Analysis*, 74:102233, 2021.

[108] Yijing Li, Xiaofeng Tao, Xuefei Zhang, Mingsi Wang, and Shuo Wang. Break the data barriers while keeping privacy: A graph differential privacy method. *IEEE Internet Things J.*, 2022.

[109] Wanyu Lin, Baochun Li, and Cong Wang. Towards private learning on decentralized graphs with local differential privacy. *IEEE Trans. Inf. Forensics Secur.*, 17:2936–2946, 2022.

[110] Bin Liu, Zhen Li, Xi Chen, Yuehui Huang, and Xiangdong Liu. Recognition and vulnerability analysis of key nodes in power grid based on complex network centrality. *IEEE Trans. Circuits and Syst. II, Express Briefs*, 65(3):346–350, 2017.

[111] Bin Liu, Wei Ni, Ren Ping Liu, Qi Zhu, Y Jay Guo, and Hongbo Zhu. Novel integrated framework of unmanned aerial vehicle and road traffic for energy-efficient delay-sensitive delivery. *IEEE Trans. Intell. Transp. Syst.s*, 23(8):10692–10707, 2021.

[112] Yueliang Liu et al. Graph learning based on spatiotemporal smoothness for time-varying graph signal. *IEEE Access*, 7:62372–62386, 2019.

[113] Zishan Liu, Lin Zhang, Wei Ni, and Iain B Collings. Uncoordinated pseudonym changes for privacy preserving in distributed networks. *IEEE Trans. Mob. Comput.*, 19(6):1465–1477, 2019.

[114] Imran Makhdoom, Ian Zhou, Mehran Abolhasan, Justin Lipman, and Wei Ni. Privysharing: A blockchain-based framework for privacy-preserving and secure data sharing in smart cities. *Comput. Secur.*, 88:101653, 2020.

[115] Guillaume Marrelec and Peter Fransson. Assessing the influence of different ROI selection strategies on functional connectivity analyses of fMRI data acquired during steady-state conditions. *PloS one*, 6(4):e14788, 2011.

[116] Gonzalo Mateos et al. Connecting the dots: Identifying network structure via graph signal processing. *IEEE Signal Process. Mag.*, 36(3):16–43, 2019.

[117] Michael E Mavroforakis and Sergios Theodoridis. A geometric approach to support vector machine (svm) classification. *IEEE Trans. Neural Netw.*, 17(3):671–682, 2006.

[118] Troy McMahon, Adrian Chan, Shlomo Havlin, and Lazaros K Gallos. Spatial correlations in geographical spreading of Covid-19 in the United States. *Sci. Rep.*, 12(1):1–10, 2022.

[119] Marianna Milano. Cctv: A new network-based methodology for the analysis and visualization of Covid-19 data. In *Proc. Int. Conf. Bioinformat. Biomed. (BIBM)*, pages 2000–2001. IEEE, 2021.

[120] Sushil Mittal and Peter Meer. Conjugate gradient on Grassmann manifolds for robust subspace estimation. *Image Vis. Comput.*, 30(6-7):417–427, 2012.

[121] Tamara T Mueller, Dmitrii Usynin, Johannes C Paetzold, Rickmer Braren, Daniel Rueckert, and Georgios Kaissis. Differentially private guarantees for analytics and machine learning on graphs: A survey of results. *Journal of Privacy and Confidentiality*, 14(1), 2024.

[122] Sunil K Narang and Antonio Ortega. Compact support biorthogonal wavelet filterbanks for arbitrary undirected graphs. *IEEE Trans. Signal Process.*, 61(19):4673–4685, 2013.

[123] ARAL Neşe and Hasan Bakir. Spatiotemporal analysis of Covid-19 in turkey. *Sustain. Cities Soc.*, 76:103421, 2022.

[124] Dinh C Nguyen, Quoc-Viet Pham, Pubudu N Pathirana, Ming Ding, Aruna Seneviratne, Zihuai Lin, Octavia Dobre, and Won-Joo Hwang.

Federated learning for smart healthcare: A survey. *ACM Computing Surveys (Csur)*, 55(3):1–37, 2022.

[125] Wei Ni, Iain B Collings, Justin Lipman, Xin Wang, Meixia Tao, and Mehran Abolhasan. Graph theory and its applications to future network planning: Software-defined online small cell management. *IEEE Wirel. Commun.*, 22(1):52–60, 2015.

[126] Ryo Okui. Asymptotically unbiased estimation of autocovariances and autocorrelations with long panel data. *Econometric Theory*, 26(5):1263–1304, 2010.

[127] Iyiola E Olatunji, Wolfgang Nejdl, and Megha Khosla. Membership inference attack on graph neural networks. In *2021 Third IEEE International Conference on Trust, Privacy and Security in Intelligent Systems and Applications (TPS-ISA)*, pages 11–20. IEEE, 2021.

[128] Antonio Ortega, Pascal Frossard, Jelena Kovačević, José MF Moura, and Pierre Vandergheynst. Graph signal processing: Overview, challenges, and applications. *Proceedings of the IEEE*, 106(5):808–828, 2018.

[129] Harry Oviedo, Oscar Dalmau, and Hugo Lara. Two adaptive scaled gradient projection methods for stiefel manifold constrained optimization. *Numerical Algorithms*, 87(3):1107–1127, 2021.

[130] Yue Pan, Limao Zhang, Juliette Unwin, and Miroslaw J Skibniewski. Discovering spatial-temporal patterns via complex networks in investigating Covid-19 pandemic in the United States. *Sustain. Cities Soc.*, 77:103508, 2022.

[131] Bastien Pasdeloup, Vincent Gripon, Grégoire Mercier, Dominique Pastor, and Michael G Rabbat. Characterization and inference of graph diffusion processes from observations of stationary signals. *IEEE Trans. Signal Inf. Process. Netw.*, 4(3):481–496, 2017.

[132] Nathanaël Perraudin and Pierre Vandergheynst. Stationary signal processing on graphs. *IEEE Trans. Signal Process.*, 65(13):3462–3477, 2017.

[133] Saiprasad Ravishankar and Yoram Bresler. Sparsifying transform learning with efficient optimal updates and convergence guarantees. *IEEE Trans. Signal Process.*, 63(9):2389–2404, 2015.

[134] Muhammad Ahmad Raza, Mehran Abolhasan, Justin Lipman, Negin Shariati, Wei Ni, and Abbas Jamalipour. Statistical learning-based adaptive network access for the industrial internet-of-things. *IEEE Internet Things J.*, 2023.

[135] Seyed Saman Saboksayr, Gonzalo Mateos, and Mujdat Cetin. Online discriminative graph learning from multi-class smooth signals. *Signal Process.*, 186:108101, 2021.

[136] Aliaksei Sandryhaila and José MF Moura. Discrete signal processing on graphs. *IEEE Trans. Signal Process.*, 61(7):1644–1656, 2013.

[137] Aliaksei Sandryhaila and Jose MF Moura. Discrete signal processing on graphs: Frequency analysis. *IEEE Trans. Signal Process.*, 62(12):3042–3054, 2014.

[138] Saeid Sanei and Jonathon A Chambers. *EEG signal processing*. John Wiley & Sons, 2013.

[139] Stefania Sardellitti, Sergio Barbarossa, and Paolo Di Lorenzo. Graph topology inference based on transform learning. In *2016 IEEE Global Conference on Signal and Information Processing (GlobalSIP)*, pages 356–360. IEEE, 2016.

[140] Stefania Sardellitti et al. Graph topology inference based on sparsifying transform learning. *IEEE Trans. Signal Process.*, 67(7):1712–1727, 2019.

[141] Anand D Sarwate and Kamalika Chaudhuri. Signal processing and machine learning with differential privacy: Algorithms and challenges for continuous data. *IEEE Signal Process. Mag.*, 30(5):86–94, 2013.

[142] Santiago Segarra, Antonio G Marques, Gonzalo Mateos, and Alejandro Ribeiro. Network topology inference from spectral templates. *IEEE Trans. Signal Inf. Process. Netw.*, 3(3):467–483, 2017.

[143] Baoling Shan, Wei Ni, Xin Yuan, Dongwen Yang, Xin Wang, and Ren Ping Liu. Graph learning from band-limited data by graph Fourier transform analysis. *Signal Process.*, 207:108950, 2023.

[144] Kaiming Shen and Wei Yu. Fractional programming for communication systems—Part I: Power control and beamforming. *IEEE Trans. Signal Process.*, 66(10):2616–2630, 2018.

[145] Hadi Shirouyehzad, Javid Jouzdani, and Mazdak Khodadadi Karimvand. Fight against Covid-19: A global efficiency evaluation based on contagion control and medical treatment. *J. appl. res. ind. eng.*, 7(2):109–120, 2020.

[146] David I Shuman, Sunil K Narang, Pascal Frossard, Antonio Ortega, and Pierre Vandergheynst. The emerging field of signal processing on graphs: Extending high-dimensional data analysis to networks and other irregular domains. *IEEE signal process. mag.*, 30(3):83–98, 2013.

[147] Mike KP So, Amanda MY Chu, Agnes Tiwari, and Jacky NL Chan. On topological properties of Covid-19: Predicting and assessing pandemic risk with network statistics. *Sci. Rep.*, 11(1):1–14, 2021.

[148] Zixing Song, Xiangli Yang, Zenglin Xu, and Irwin King. Graph-based semi-supervised learning: A comprehensive review. *IEEE Transactions on Neural Networks and Learning Systems*, 34(11):8174–8194, 2023.

[149] Ljubiša Stanković, Danilo Mandic, Miloš Daković, Miloš Brajović, Bruno Scalzo, Shengxi Li, Anthony G Constantinides, et al. Data analytics on graphs part i: Graphs and spectra on graphs. *Foundations and Trends® in Machine Learning*, 13(1):1–157, 2020.

[150] Housheng Su, Zhihai Rong, Michael ZQ Chen, Xiaofan Wang, Guanrong Chen, and Hongwei Wang. Decentralized adaptive pinning control for cluster synchronization of complex dynamical networks. *IEEE Trans. Cybern.*, 43(1):394–399, 2012.

[151] Simon Syga, Diana David-Rus, Yannik Schälte, Haralampos Hatzikirou, and Andreas Deutsch. Inferring the effect of interventions on Covid-19 transmission networks. *Sci. Rep.*, 11(1):1–11, 2021.

[152] Xiwei Tang, Jianxin Wang, Jiancheng Zhong, and Yi Pan. Predicting essential proteins based on weighted degree centrality. *IEEE/ACM Trans. Comput. Biol. Bioinf.*, 11(2):407–418, 2013.

[153] Dimitrios Tsiotas and Vassilis Tselios. Understanding the uneven spread of Covid-19 in the context of the global interconnected economy. *Sci. Rep.*, 12(1):1–15, 2022.

[154] Mikhail Tsitsvero, Sergio Barbarossa, and Paolo Di Lorenzo. Signals on graphs: Uncertainty principle and sampling. *IEEE Trans. Signal Process.*, 64(18):4845–4860, 2016.

[155] Nathalie Tzourio-Mazoyer, Brigitte Landeau, Dimitri Papathanassiou, Fabrice Crivello, Octave Etard, Nicolas Delcroix, Bernard Mazoyer, and Marc Joliot. Automated anatomical labeling of activations in SPM using a macroscopic anatomical parcellation of the MNI MRI single-subject brain. *Neuroimage*, 15(1):273–289, 2002.

[156] Hadrien Van Lierde, Tommy WS Chow, and Guanrong Chen. Scalable spectral clustering for overlapping community detection in large-scale networks. *IEEE Trans. Knowl. Data Eng.*, 32(4):754–767, 2019.

[157] Jian Wang and Renguang Zuo. Assessing geochemical anomalies using geographically weighted lasso. *Applied Geochemistry*, 119:104668, 2020.

[158] Songlei Wang, Yifeng Zheng, Xiaohua Jia, and Xun Yi. PeGraph: A system for privacy-preserving and efficient search over encrypted social graphs. *IEEE Trans. Inf. Forensics Secur.*, 17:3179–3194, 2022.

[159] Xinyuan Wang, Yingze Wang, Qimei Cui, Kwang-Cheng Chen, and Wei Ni. Machine learning enables radio resource allocation in the downlink of ultra-low latency vehicular networks. *IEEE Access*, 10:44710–44723, 2022.

[160] Yue Wang and Xintao Wu. Preserving differential privacy in degree-correlation based graph generation. *Transactions on data privacy*, 6(2):127, 2013.

[161] Yulong Wang, Tianxiang Li, Shenghong Li, Xin Yuan, and Wei Ni. New adversarial image detection based on sentiment analysis. *IEEE Internet Things J.*, 2023.

[162] Yulong Wang, Tong Sun, Shenghong Li, Xin Yuan, Wei Ni, Ekram Hossain, and H Vincent Poor. Adversarial attacks and defenses in machine learning-empowered communication systems and networks: A contemporary survey. *IEEE Communications Surveys & Tutorials*, 2023.

[163] Zhongyu Wang, Zhipeng Lin, Tiejun Lv, and Wei Ni. Energy-efficient resource allocation in massive MIMO-NOMA networks with wireless power transfer: A distributed admm approach. *IEEE Internet Things J.*, 8(18):14232–14247, 2021.

[164] Chengkun Wei, Shouling Ji, Changchang Liu, Wenzhi Chen, and Ting Wang. AsgLDP: collecting and generating decentralized attributed graphs with local differential privacy. *IEEE Trans. Inf. Forensics Secur.*, 15:3239–3254, 2020.

[165] World Health Organization. Coronavirus disease (Covid-19) pandemic.

[166] World Health Organization. Coronavirus disease (Covid-19): Variants of Sars-Cov-2.

[167] Zonghan Wu, Shirui Pan, Fengwen Chen, Guodong Long, Chengqi Zhang, and S Yu Philip. A comprehensive survey on graph neural networks. *IEEE transactions on neural networks and learning systems*, 32(1):4–24, 2020.

[168] Mingrui Xia, Jinhui Wang, and Yong He. Brainnet viewer: A network visualization tool for human brain connectomics. *PloS one*, 8(7):e68910, 2013.

[169] Keyulu Xu, Weihua Hu, Jure Leskovec, and Stefanie Jegelka. How powerful are graph neural networks? *arXiv preprint arXiv:1810.00826*, 2018.

[170] Dongwen Yang, Wei Ni, Lan Du, Hongwei Liu, and Jiadong Wang. Efficient attributed scatter center extraction based on image-domain sparse representation. *IEEE Trans. Signal Process.*, 68:4368–4381, 2020.

[171] Sen Yang, Qian Sun, Shuiwang Ji, Peter Wonka, Ian Davidson, and Jieping Ye. Structural graphical lasso for learning mouse brain connectivity. In *Proceedings of the 21st ACM SIGKDD International Conference on Knowledge Discovery and Data Mining*, pages 1385–1394, 2015.

[172] Chia-Chen Yen, Mi-Yen Yeh, and Ming-Syan Chen. An efficient approach to updating closeness centrality and average path length in dynamic networks. In *2013 IEEE 13th International Conference on Data Mining*, pages 867–876. IEEE, 2013.

[173] Feiran You, Xin Yuan, Wei Ni, and Abbas Jamalipour. A novel privacy-preserving incentive mechanism for multi-access edge computing. *IEEE Trans. Cogn. Commun. Netw.*, pages 1–1, 2024.

[174] Fahong Yu, Meijia Chen, Bolin Yu, Wenping Li, Longhua Ma, and Huimin Gao. Privacy preservation based on clustering perturbation algorithm for social network. *Multimedia Tools and Applications*, 77(9):11241–11258, 2018.

[175] Mingxuan Yuan, Lei Chen, S Yu Philip, and Ting Yu. Protecting sensitive labels in social network data anonymization. *IEEE Trans. Knowl. Data Eng.*, 25(3):633–647, 2011.

[176] Xin Yuan, Zhiyong Feng, J Andrew Zhang, Wei Ni, Ren Ping Liu, Zhiqing Wei, and Changqiao Xu. Spatio-temporal power optimization for mimo joint communication and radio sensing systems with training overhead. *IEEE Trans. Veh. Technol.*, 70(1):514–528, 2020.

[177] Kun Zhan, Changqing Zhang, Junpeng Guan, and Junsheng Wang. Graph learning for multiview clustering. *IEEE Trans. Cybern.*, 48(10):2887–2895, 2017.

[178] Jianjia Zhang, Luping Zhou, and Lei Wang. Subject-adaptive integration of multiple SICE brain networks with different sparsity. *Pattern Recognit.*, 63:642–652, 2017.

[179] K. Zhang, Y. Zhang, R. Sun, P. Tsai, M. Ul Hassan, X. Yuan, M. Xue, and J. Chen. Bounded and unbiased composite differential privacy. In *2024 IEEE Symposium on Security and Privacy (SP)*, pages 111–111, Los Alamitos, CA, USA, may 2024. IEEE Computer Society.

[180] Jun Zhao, Teng Wang, Tao Bai, Kwok-Yan Lam, Zhiying Xu, Shuyu Shi, Xuebin Ren, Xinyu Yang, Yang Liu, and Han Yu. Reviewing and improving the gaussian mechanism for differential privacy. *arXiv preprint arXiv:1911.12060*, 2019.

[181] Kanhao Zhao, Boris Duka, Hua Xie, Desmond J Oathes, Vince Calhoun, and Yu Zhang. A dynamic graph convolutional neural network framework reveals new insights into connectome dysfunctions in ADHD. *NeuroImage*, 246:118774, 2022.

[182] Licheng Zhao et al. Optimization algorithms for graph Laplacian estimation via ADMM and MM. *IEEE Trans. Signal Process.*, 67(16):4231–4244, 2019.

[183] Jingjing Zheng, Xin Yuan, Kai Li, Wei Ni, Eduardo Tovar, and Jon Crowcroft. A novel defense against poisoning attacks on federated learning: Layercam augmented with autoencoder. *arXiv preprint arXiv:2406.02605*, 2024.

[184] Houliang Zhou, Lifang He, Yu Zhang, Li Shen, and Brian Chen. Interpretable graph convolutional network of multi-modality brain imaging for alzheimer's disease diagnosis. In *2022 IEEE 19th International Symposium on Biomedical Imaging (ISBI)*, pages 1–5. IEEE, 2022.

[185] Jie Zhou, Ganqu Cui, Shengding Hu, Zhengyan Zhang, Cheng Yang, Zhiyuan Liu, Lifeng Wang, Changcheng Li, and Maosong Sun. Graph neural networks: A review of methods and applications. *AI open*, 1:57–81, 2020.

[186] Jingya Zhou, Ling Liu, Wenqi Wei, and Jianxi Fan. Network representation learning: from preprocessing, feature extraction to node embedding. *ACM Comput. Surv.*, 55(2):1–35, 2022.

[187] Yiming Zuo, Yi Cui, Guoqiang Yu, Ruijiang Li, and Habtom W Ressom. Incorporating prior biological knowledge for network-based differential gene expression analysis using differentially weighted graphical lasso. *BMC bioinformatics*, 18:1–14, 2017.

Index

Note: **Bold** page numbers refer to **tables** and *Italic* page numbers refer to *figures*.

Printed in the United States
by Baker & Taylor Publisher Services